Carbon Membrane
Technology

Carbon Membrane Technology

Fundamentals and Applications

Edited By

Xuezhong He and Izumi Kumakiri

CRC Press
Taylor & Francis Group
Boca Raton London New York

CRC Press is an imprint of the
Taylor & Francis Group, an **informa** business

First edition published 2021
by CRC Press
6000 Broken Sound Parkway NW, Suite 300, Boca Raton, FL 33487-2742

and by CRC Press
2 Park Square, Milton Park, Abingdon, Oxon, OX14 4RN

Library of Congress Cataloging-in-Publication Data

ISBN: 978-1-138-33337-6 (hbk)
ISBN: 978-0-429-44598-9 (ebk)

Typeset in Palatino
by SPi Global, India

Contents

Preface

Gas and liquid separation by membrane technology has become a topic of rapidly growing interest as an alternative to traditional separation methods within the past 30 years. The membrane technique is a low-cost, energy-efficient process that requires simple, compact, and easily operated equipment. Moreover, there is no requirement for solvents and other chemicals for this environmentally friendly process. However, the biggest challenges with membrane technology are the requirement to pretreat the feed gas stream, and the membrane lifetime. Carbon membranes have great advantages in terms of mechanical and chemical stability and high separation performance, which have allowed their development in the past two decades as promising candidates for energy-efficient gas purification processes such as H_2/CO_2 and CO_2/CH_4 separation. Further improvements in membrane performance have the potential to offset their relatively high production cost compared with polymeric membranes. However, there are still some challenges related to making asymmetric carbon membranes, controlling the structure and pore size, and module upscaling for commercial applications. The aim of this book is to provide the fundamentals on carbon membrane materials for young researchers developing frontier membranes for energy-efficient separation processes.

This book describes both self-supported and supported carbon membranes, from fundamentals to applications. The key steps in the development of high-performance carbon membranes, including precursor selection, tuning carbon membrane structure, and regeneration, are discussed. Different characterization methods are used to analyze the structure, morphology, and performance. The transport mechanisms are analyzed experimentally and via computational simulation. Finally, potential applications in both gas and liquid separation are described, and the future directions for carbon membrane development pointed out.

Membrane science and engineering are set to play crucial roles enabling technologies to provide energy-efficient and cost-effective future solutions for energy and environment-related processes. Based on this approach, many current research projects are trying to find attractive carbon materials. Published papers on the topic of carbon membranes, especially for biogas upgrading, natural gas sweetening, and hydrogen purification, are numerous and have very high impact. However, there are few books relevant to the topic of carbon membrane technology. This book will receive a warm welcome from postgraduate students and industrial engineers. The knowledge offered in this book is condensed and interdisciplinary at the same time.

The authors of this book intend to make a tutorial overview of the major types of carbon membranes and to provide information concerning the main applications in gas and liquid separation and future perspectives on carbon membranes from materials to processes.

This book consists of 10 chapters and has two main parts. In the first part, four chapters refer to the fundamentals of carbon membrane development, characterization, and transport mechanisms. The second part discusses carbon membranes for different applications such as biogas upgrading, natural gas sweetening, and liquid separation related to microfiltration/ultrafiltration/nanofiltration processes.

To further analyze this book, *"Part 1. Fundamentals of Carbon Membranes"* starts with Chapter 1 (Lei and He), which provides an overview of carbon membrane preparation and regeneration. Chapter 2 (Lei and He) provides an overview of the separation performances of different types of carbon membranes such as self-supported hollow fibers, supported flat sheets, and those supported on ceramic tubes. Chapter 3 (Lei) describes different methods such as thermogravimetric analysis–mass spectrometry (TGA–MS), dynamic mechanical analysis, scanning electron microscopy (SEM), X-ray diffraction (XRD), and gas sorption and permeation for characterization of carbon membrane structure, mechanical strength, and separation performance. Chapter 4 (He) describes carbon membranes used for gas separation based on different transport mechanisms including molecular sieving, selective surface flow, and Knudsen diffusion. Moreover, the general mass transport through membrane and process parameter influences are discussed, which will be used to guide membrane modeling and process simulation for specific applications of carbon membranes.

"Part 2. Carbon Membrane Applications" focuses on carbon membranes for different gas or liquid separations. Chapter 5 (Lindbråthen) reports the experiences gained from running a small pilot-scale carbon membrane upgrading pilot on a site in Norway. The municipal- and food-waste-generated biogas was ultimately successfully upgraded to biomethane quality and used as fuel for cars. The chapter also discuss the pitfalls and experiences gained in the construction of the pilot carbon membrane system for biogas upgrading. Chapter 6 (Favvas, Kaldis, and He) reviews the main steps and achievements of carbon membranes in natural gas sweetening processes. Chapter 7 (Yoshimune) focuses on carbon membranes for use in hydrogen (H_2) purification. These processes include the separation of gases such as H_2/nitrogen (N_2), H_2/methane (CH_4), H_2/carbon dioxide (CO_2), and H_2/hydrocarbons. Chapter 8 (Zhang and Wu) touches on carbon membranes for liquid separation, especially microfiltration, ultrafiltration, and nanofiltration. Chapter 9 (Gomez, Kumakiri, and He) describes other applications of carbon membranes for gas separation, such as CO_2 capture from flue gas and olefin/paraffin separation in petrochemical industry. Chapter 10 (He) highlights the future research directions and engineering requirements with

respect to green advances in carbon membrane material development as well as sustainable solutions of carbon membranes for industrial separation processes.

With this book, CRC Press (Taylor & Francis Group) gives an opportunity to the scientists who are working in the field of carbon membrane technology for gas and liquid separations to present, share, and discuss with the membrane community their contribution. At this point, both editors enthusiastically acknowledge all those who by different ways contributed to this book, including those at CRC Press and all the chapters' authors.

The strong support given by Dr. Gagandeep Singh (Senior Editor in Engineering & Environmental Sciences at CRC Press) has been also greatly appreciated.

Xuezhong He
Izumi Kumakiri

Editors

Xuezhong He is an associate professor at Guangdong Technion Israel Institute of Technology (GTIIT) and Technion-Israel Institute of Technology, who has been working in the Department of Chemical Engineering at Norwegian University of Science and Technology (NTNU) for more than 10 years in membrane science and engineering. Dr. He has solid knowledge of and high competence in membrane materials (both polymeric and carbon membranes in flat sheets and hollow fibers) for energy-related separation processes, especially natural gas sweetening, biogas upgrading, and hydrogen purification. Dr. He has published >40 articles in peer-reviewed journals such as *J. Member. Sci.*, *Chem. Eng. J.*, *AIChE J.*, *I&EC Res.*, and has >2000 citations and an H-index of 20 (Google Scholar), and has contributed eight chapters to books; Dr. He has also given a number of presentations at international conferences.

Izumi Kumakiri is from Yamaguchi University Japan, having previously worked at National Center for Scientific Research (CNRS), France, and at Stiftelsen for industriell og teknisk forskning (SINTEF), Norway, and is involved in various national, international, and industrial projects on the developments and the applications of membranes and nano-functionalized materials. Professor Kumakiri is currently conducting research projects related to the developments of microporous inorganic membranes including carbon, zeolite, and other types, and their applications in energy and environmental fields, and has published more than 50 articles in leading scientific journals.

Part 1

Fundamentals of Carbon Membranes

1

Carbon Membrane Preparation

Linfeng Lei[a] and Xuezhong He[a,b]

[a]*Department of Chemical Engineering, Norwegian University of Science and Technology*

[b]*Department of Chemical Engineering, Guangdong Technion Israel Institute of Technology (GTIIT)*

1.1 Introduction

Carbon membranes, which have strong mechanical and chemical stability and high separation performance, have been developed in recent decades as promising candidates for energy-efficient gas purification processes such as H_2/CO_2 [1], CO_2/N_2 [2], and CO_2/CH_4 separation [3–5]. Since the first carbon membranes were produced by carbonization of cellulose hollow fibers [6], carbon membranes developed from precursors such as cellulose derivatives [7–12], polyimide derivatives [13–19], poly(vinylidene fluoride) [20], and polyacrylonitrile [21] have been investigated systematically for different applications (CO_2/CH_4 and olefin/paraffin separation [13, 22, 23]) and have demonstrated excellent separation performances. Moreover, by carefully controlling the carbonization conditions (e.g., final temperature, heating rate, and environment) and introducing proper post treatment, the pore structure and porosity of prepared carbon membranes can be tailored to separate gas molecules that are alike in both size and physical properties [13, 23, 24]. Challenging separations, such as H_2/CO_2 and olefin/paraffin, can be addressed by pore tailoring—using different purge gases (e.g., CO_2, N_2), final carbonization temperatures, and soak times—and post treatment (oxidation and chemical vapor deposition) as well as tuning the precursor structure.

1.2 Precursor Selection and Preparation

In past decades, various aromatic polymeric materials have been used for the fabrication of high-performance carbon membranes, including cellulose and derivatives [11, 12, 25, 26], polyimide and derivatives [19, 27–30], polyacrylonitrile [21, 31], poly(*p*-phenylene oxide) [32–34], and phenolic resin [35, 36], as listed in Table 1.1. Generally, a suitable polymeric precursor material for producing a carbon membrane should not cause pore holes or any defects after carbonization, as the separation is based on a molecular sieving transport mechanism. Moreover, precursor-determined properties, such as chemical structure, glass transition temperature, and fractional free volume (FFV), should be considered when selecting ideal precursors for carbon membranes [37, 38]. For example, Park et al. [37] found that adding methyl substituent groups, which increase the FFV, to a polyimide backbone can enhance permeability. Polyimides and derivatives are the most-used polymers for making carbon molecular sieve (CMS) membranes, due to their thermal stability and properties of maintaining morphology, but they are often expensive, which might limit their wide application. On the other hand, cellulose, as a sustainable raw material with relatively low cost, shows great potential for the fabrication of carbon membranes [12, 39].

To date, *N*-methyl-2-pyrrolidone is the most common solvent for polymer dope solutions and subsequent preparation of precursors. Employing ionic liquid solvents for precursor fabrication has emerged as a promising alternative to approach a green and clean chemistry. 1-ethyl-3-methylimidazolium acetate (EmimAc), a commercial product, has been successfully used for the fabrication of cellulose membrane precursors and the production of CMS membranes [11, 12]. Rodrigues et al. [12] reported a dense CMS film that surpassed the Robeson upper bound [40] of O_2/N_2, He/N_2, and H_2/N_2 and that was derived from EmimAc-regenerated cellulose precursors. Owing to the highly hydrophilic characteristic that allows water vapor to permeate quickly, the prepared CMS membranes exhibited good humidity stability in the presence of 75–77 % relative humidity at 25 °C.

For hollow fiber precursor preparation, a typical dry–wet spinning process with two coagulation baths was employed, as illustrated in Figure 1.1. The dope solution, which consists of polymer and solvent (and also additives such as nonsolvent, metal, or polymer additives), is pumped through a spinneret of a specific dimension while bore fluid is injected simultaneously through the spinneret's inner tube. After a short residence time in the air (or a specific environment such as a humidified atmosphere), the nascent hollow fiber reaches a nonsolvent coagulation bath and is stretched by the take-up wheel to form a solidified hollow fiber. The spinning parameters—length of air gap, take-up speed, bore and dope fluid flow rate, temperature of dope solution and coagulation bath—have significant effects on membrane morphology and the structure and properties of the hollow fibers.

Peng et al. [41] summarized some key factors in the hollow fiber spinning process. For example, a higher polymer concentration in the dope solution will result in stronger chain entanglement, which can effectively lower macrovoids and porosity in the membrane matrix. The coagulation rate (enhanced by a higher solubility difference) can also change the morphology

TABLE 1.1

Typical precursor materials for CMS membranes.

Precursor material	Configuration	Carbonization conditions (temperature, purge gas)	References
Cellulose	Flat sheet	500–850 °C, mild vacuum (0.5 mbar)	[7]
	Flat sheet	550–600 °C, N_2	[12]
	Hollow fiber	500–800 °C, Ar	[61]
Cellulose acetate	Hollow fiber	550 °C, CO_2	[9, 25, 26]
	Hollow fiber	550–750 °C, N_2, CO_2, vacuum	[10]
	Hollow fiber	650 °C, CO_2	[8]
	Hollow fiber	550 °C, N_2	[62, 63]
Polyacrylonitrile	Hollow fiber	600–950 °C, N_2	[31]
		500 °C, N_2	[64]
		900 °C, N_2	[65]
Polyethyleneimine	Supported film	600 °C, –	[66]
	Supported film	600 °C, –	[67]
Polyimide	Hollow fiber	500–800 °C, Ar	[27, 28]
	Hollow fiber	750–900 °C, Ar	[19]
	Hollow fiber	550 °C, Ar	[29, 52, 68]
	Hollow fiber	500–550 °C, Ar	[69]
	Hollow fiber and film	550–800 °C, Ar	[42, 70]
	Film	500–800 °C, Ar	[71, 72]
	Film	800 °C, vacuum	[73]
	Supported film	700 °C, vacuum	[1]
Polyphenylene oxide	Hollow fiber	550–750 °C, vacuum	[33]
	Film	400–700 °C, vacuum	[34]
Phenol resin	Film	700–850 °C, N_2, CO_2	[74]
	Supported film	700 °C, vacuum	[48, 75, 76]
	Supported film	800 °C, N_2	[77]
	Supported film	600–900 °C, vacuum	[78]

of the spun hollow fibers. Generally, a low coagulation rate will induce a thick and relatively porous skin whereas a thinner but relatively dense layer is created with faster coagulation. Because the coagulation of nascent hollow fibers at the inner side, which reacts with the bore fluid, and the outer side, which reacts with the coagulation bath, take place simultaneously, the

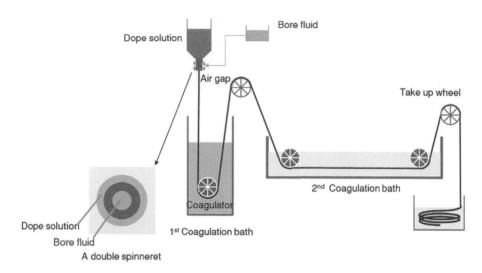

Dope solution
Bore fluid
Air gap
Take up wheel
2nd Coagulation bath
Coagulator
Dope solution
Bore fluid
A double spinneret
1st Coagulation bath

FIGURE 1.1
Schematic of a typical dry–wet spinning process using a double spinneret.

membrane morphology can be adjusted by controlling the composition of the nonsolvents.

Lei et al. [11] investigated the influence of spinning parameters on the separation performance of carbon hollow fibers. They found that CO_2 permeability was significantly improved by increasing the air gap and dope solution flow rate but weakened with increased water content in the bore fluid. Concomitantly, CO_2/CH_4 selectivity was enhanced with elevated dope solution flow rate but decreased with an increase in bore flow and take-up speed. This prepared precursor structure and property will have a crucial influence on the performance of the derived carbon membrane.

The precursor membranes are often pretreated before being carbonized, because this can help to enhance the stability of the precursor during carbonization (e.g., avoid pore collapse). Thus, the performance of CMS membranes can be improved by introducing a specific pretreatment. The pretreatment approaches can be divided into physical and chemical methods. Physical pretreatment is often used during the spinning process to eliminate surface defects and enhance the retention of molecular orientation. Chemical pretreatments involve the application of chemical reagents to the polymeric precursors. A precursor may be subjected to more than one pretreatment method to achieve the desired properties in the carbon membrane. For example, an organic silica that creates an extra layer of the polyimide hollow fiber precursor to avoid pore collapse during the carbonization process has been utilized [16, 22, 42].

1.3 Carbon Membrane Preparation

During carbonization (see the illustration in Figure 1.2), the entangled polymeric precursor is transformed to rigidly carbonized aromatic strands and, afterwards, forms organized plates. As a result, the carbon membranes present a bimodal structure with ultramicropores and micropores. Although carbon membranes can be prepared from different precursors, which are likely to form different final carbon structures and thus give different separation performances, the general evolution from precursor to carbon during carbonization is supposed to be similar. Rungta et al. [22] visualized the formation of organized CMS membranes from polyimide during carbonization, as illustrated in Figure 1.3. Initially, the entangled precursor is activated to undergo aromatization and fragmentation during the temperature ramping process, then periodic scissions are generated along the polymer backbone (Figure 1.3(i and ii)) [22, 43]. The backbone scissions are transformed to rigidly aromatic carbon strands after most of the oxygen atoms are removed [43], then the carbon strands align to form plates, reaching higher entropy and reducing the excluded volume effects that present within randomly packed strands (Figure 1.3(iii)) [22]. However, it is hard to form perfect long-range stacks of plates during the final thermal soaking phase because of kinetic restrictions. Thus, a carbon membrane presents an idealized micropore with imperfectly

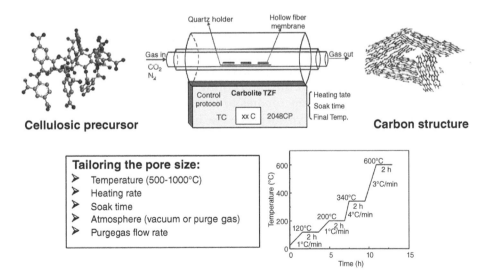

FIGURE 1.2
Illustration of a carbonization process.

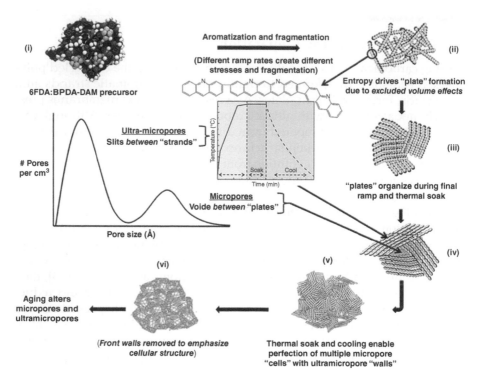

FIGURE 1.3
Envisioned initial steps in transformation from random precursor polyimide to organized amorphous CMS material with bimodal distribution of micropores and ultramicropores [22].

packed plates that is formed by the organized strands (ultramicropore), as depicted in Figure 1.3(iv). At the end cooling stage, the ultramicropore will share the "walls" between micropores to form a cellular structure, as illustrated in Figure 1.3(v, vi). A typical pore-size distribution of CMS membranes measured by CO_2 adsorption at 0 °C is shown in Figure 1.4, which provides clear proof of the envisioned CMS structures.

Normally, the pore size of ultramicropores is ca. 3–7 Å, which is suggested to govern gas-pair selectivity while the micropores (7–20 Å) enhance gas permeance [18, 19, 43]. The carbon membrane has a sp^2- and sp^3-hybridized carbon structure [19, 43], in which the sp^2-hybridized carbon, a two-dimensional layered graphitic carbon, is beneficial to plate packing that results in a more compact ultramicropore structure. The sp^3-hybridized carbon, as a three-dimensional structure, is obstructive to plate packing, leading to wider micropores and thus inducing higher gas permeance. The sp^3-hybridized carbon structure is thermodynamically unstable, so it can be partly transformed to an sp^2 structure under appropriate conditions [39, 44]. Thus, tuning the carbonization conditions such as temperature and atmosphere could

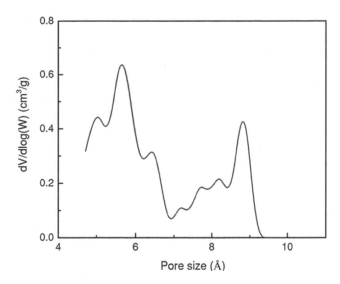

FIGURE 1.4
A typical pore-size distribution of a CMS membrane, measured by CO_2 adsorption at 0 °C [11].

provide some facile ways to modify CMS structures and thus improve separation performance.

Recently, Ma et al. reported a H_2-assisted approach to create well-defined "mid-sized" micropores in CMS membranes during the carbonization process [43]. The introduction of H_2 to the carbonization environment can increase the size of the ultramicropores but retain the size of the micropores, and ultimately lead to a dramatic improvement in separation permeability with little loss of selectivity. H_2 in the carbonization environment can reduce the formation of CO_2 but promote the loss of oxygen in the form of H_2O. The production of H_2O and CO_2 has been shown by off-gas mass spectrometry. The authors also found that the pore size of CMS membranes can be tuned by adjusting the H_2 concentration in an Ar atmosphere [43]. Specifically, CMS membranes carbonized with the increasing amounts of H_2 (0, 1, 2, and 4 vol% in Ar) present correspondingly increased pore sizes. Moreover, a carbonization environment that contains H_2 can modulate the carbon hybridized structures (sp^2 and sp^3). It was reported that the sp^3/sp^2 ratio is increased with increased H_2 concentration, which results in a more permeable but less selective structure [43].

Oxidative treatments to widen the pores in the carbon membrane have been reported [45–48]. For example, Richter et al. [46] showed that carbon membranes exposed to air at different temperatures (e.g., 300, 350, and 400 °C) have different degrees of gas permeance. The H_2 permeances were increased from 5 to 18 m^3 (standard temperature and pressure)/($m^2 \cdot h \cdot bar$) when exposed at 350 °C, compared with the fresh carbon membrane. Moreover, it was found that the molecular sieve carbon structure had been transformed into a carbon structure with selective surface flow, which could dramatically steer CO_2 selection in the context of CO_2/H_2 separation [46].

Increasing pore size by oxidation treatment was verified by N_2 adsorption, reported by Lee et al. [45]. Compared with the fresh carbon membranes derived from poly(phenylene oxide), the average pore size of the post-oxidized membranes increased from 0.53 to 0.68 nm when treated at 400 °C under air. As a result, the main transport mechanism was changed to Knudsen-like diffusion, which significantly deteriorated the gas selectivity.

It should be noted that introducing trace amount of O_2 at a high temperature can tighten the pore size, creating more selective CMS membranes [22, 49, 50]. It is suggested that O_2 can react with the carbon matrix and bind to the active sites, thereby narrowing the ultramicropores [49]. Gas selectivity increased with increasing O_2 content in the Ar atmosphere within a range of 4–30 ppm O_2. Nevertheless, both permeability and selectivity were reduced when O_2 content was increased to 50 ppm. The high content O_2 may cover most of the active sites of the ultramicropores and thus result in a dramatic exacerbation of transport resistance [50]. Introducing other doping species to reduce the pore size, such as ozone [51] and amine [52], has been reported elsewhere. For example, Huang et al. [51] reported an ozone-treatment-based post-synthetic modification method at room temperature to enhance separation selectivity of ultrathin CMS membranes. Compared with the untreated CMS films, the ideal gas selectivities of H_2/CH_4 and H_2/CO_2 increased from 13.3 to 50.7 and from 1.8 to 7.1, respectively, when treated by ozone. Moreover, the authors investigated the shrinkage of pore size by applying an *in situ* pulsed ozone treatment, which verified that the first single-pulse treatment reduced CH_4 transport and the subsequent pulses narrowed the pores for CO_2 transport.

Similarly, Wenz and Koros extended the doping concept by applying a paraphenylenediamine dopant [52]. They hypothesized that primary amines can react with adjacent CMS sheets to form new covalent bonds and thus decrease the ultramicropores. That work provides a new approach to tuning the transport properties of CMS membranes by introducing organic molecules whose size fits the CMS pores.

Haider et al. integrated a series of post treatments to modify separation performance, involving oxidation, reduction, and chemical vapor deposition (CVD) [47]. The authors suggested that the post-oxidized membranes were highly activated and might exhibit rapid clogging if contaminated by water or any other hydrogen-bonding molecules. Thus, post reduction conducted by flowing H_2 at 500 °C can deactivate the membrane surface but extend the micropores further. Afterwards, a new carbon layer on the CMS membrane was generated by CVD using propene at 500 °C to narrow the micropores.

The optimized carbon hollow fiber membrane (CHFM) presents 50,000 times higher CO_2 permeance and 41 times higher CO_2/CH_4 selectivity compared with the original carbon membranes.

Chemical aging in CMS membranes that is related to a sorption-induced interaction between CMS membranes and external species (e.g., water, oxygen, and organic molecules) can also reduce the pore size. It is similar to the species doping mechanism. In contrast, physical aging, which is a time-dependent process, is caused by the relaxation of the carbon "plates," and thus results in the pores shrinking. By fine control of the aging process, the membrane separation performance can be optimized; especially for H_2-related separations, as the smaller kinetic diameter of H_2 means its transport resistance through the membrane can be ignored while larger molecules like hydrocarbons are highly rejected by the reduced pores. Recently, Qiu et al. [18] reported an expedited physical aging process, named "hyperaging," to improve H_2/C_2H_4 selectivity. When the CHFMs were hyperaged in a hot flow of air with a temperature between 90 °C and 250 °C, the distance of adjacent carbon strands was compressed, thereby resulting in smaller ultramicropores. As a result, the selectivity of H_2/C_2H_4 increased more than 10-fold.

Tuning carbonization temperature is another way to control CMS membrane structure that has been reported by many researchers [7, 10, 19, 27, 53, 54]. A hypothetical evolution structure of CMS membrane with increased carbonization temperature was suggested by Zhang and Koros [19], who classified the micropores of CMS membranes into three types based on the gases that potentially can penetrate. Type I micropores are formed at relatively low carbonization temperatures and surrounded by wider ultramicropores: they can provide sorption for all tested gases, including CO_2, O_2, N_2, and CH_4. Formed at a higher carbonization temperature, type II micropores are surrounded by more refined ultramicropores. In comparison with the type I structure, type II provides sorption sites for CO_2, O_2, and N_2 but rejects CH_4 molecules. As a result, the sorption selectivities of CO_2/CH_4 and N_2/CH_4 are enhanced. If the carbonization temperature is increased further, type III micropores are generated, which are proposed to be connected by the most refined ultramicropores. Therefore, most larger gas molecules, like CH_4, N_2, and O_2, are excluded while CO_2 is allowed for sorption. Because of the narrowed pathways, both the sorption and diffusion selectivity of gas pairs can be improved. For example, a CO_2/CH_4 selectivity of above 3000 at a carbonization temperature of 900 °C was reported, which provides extremely high permselectivity for CO_2 removal from natural gas.

It can be seen that the characteristics of the microstructure can be adjusted either during carbonization or by an extra post-treatment process, which makes carbon membranes flexible in different separation applications. The pore-size-controlling approach is applicable to all membrane configurations such as self-standing flat-sheet films, hollow fibers, and supported tubulars.

1.4 Carbon Membrane Regeneration

Carbon membranes present high thermal and chemical stability, and good separation performance, but may have durability problems related to oxygen chemisorption and water condensation inside the pore structure [55]. It should be noted that a small change in pore size dramatically influences the gas permeance. Therefore, carbon membrane aging should be well considered, and regeneration methods should be applied to recover the membrane performance over time. Possible regeneration methods include:

   i. Thermal
   ii. Thermochemical
   iii. Chemical
   iv. Electrothermal

Thermal regeneration involves heating the module with an inert gas, to desorb gases or vapors from the pores and voids. Thermochemical regeneration means heating in a reactive atmosphere like air, to increase the actual pore size distribution. Reactive surface functional groups can be made passive by reduction in, for example, H_2. This method can restore or even increase the capacity of a used membrane module compared with its initial value and offers a fast route to a new module (compared with a polymeric module, which in most cases cannot be reused). The chemical method exposes the module to an agent such as propylene [56] that will dissolve some of the adsorbed gases or vapors and recover the carbon membrane porosity. Finally, electrothermal regeneration [7] releases the adsorbed molecules by applying a low-voltage current across the carbon. This requires the carbon matrix to have a certain conductivity; membranes doped with metals are suitable. For safety reasons this should be performed in a non-oxidizing gas stream.

1.5 Carbon Membrane Upscaling

Although carbon membranes have shown excellent separation performances for different applications, most of them are reported at laboratory scale. Owing to their fragility and brittleness characteristics, CMS membranes should be fabricated in free-standing hollow fibers and supported tubular configurations if considering a large scale. The membrane materials should also be considered since they relate directly to the product cost. A pilot-scale module that contains 24 medium-sized carbon hollow fiber modules (0.5–2 m² each) has been reported by Haider et al. [57]. Their carbon membranes were produced from deacetylated cellulose acetate (CA). However, the high production cost with the extra cellulose regeneration step involved, and the difficulty of keeping fibers straight during drying after CA deacetylation, are challenging the further development and applications of CA-based carbon membranes.

This chapter discusses carbon membranes produced in a batch-wise manner (except for the spinning of fibers) on a pilot scale. Some thoughts about fabrication in a continuous way are necessary. A general challenge for batch-wise production is obtaining equal quality for each membrane inside a chamber or container. The parameters involved are usually the concentrations of different compounds, and temperature. Karvan et al. [58] reported that polyimide hollow fiber precursors came in contact with each other during carbonization and fused together at high temperatures. As a result, it is difficult to avoid defects on the membrane surface. However, cellulose hollow fibers are less prone to fusing, and large quantity of cellulose fibers can be carbonized in the same batch. Haider et al. [59] reported that 1600–4000 cellulose hollow fibers were carbonized simultaneously by using 2-m-long perforated plates with square openings. They also suggested that draining the tars and vapors that were generated during carbonization is crucial to produce uniform carbon membranes.

Membrane module design and construction for high-temperature/pressure applications is another challenge when upscaling carbon membranes, related to membrane mounting, potting, and sealing. Figure 1.5 (left) illustrates the upscaling of cellulose-based CHFM modules from laboratory scale

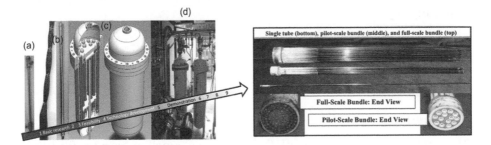

FIGURE 1.5
Upscaling of carbon membrane modules. Left: carbon hollow fiber membranes (a) laboratory-scale module, (b) medium-sized module, (c) multimodule, (d) membrane pilot plant [57]; right: tubular supported carbon membranes [60].

to a high technology readiness level; the modules have been tested success-fully for biogas upgrading. For supported carbon membranes, a membrane module consisting of tubular carbon membranes with a surface area of 0.76 m^2 and a packing density of 222 m^2/m^3 was demonstrated by Parsley et al. [60] (see Figure 1.5 (right)). A bundle of supports was fabricated before the application of carbon layers via dip-coating and carbonization. It is expected that large-scale production of supported carbon membranes will be challeng-ing, and the production cost is still quite high, which may limit the applica-tions to small-volume gas separation processes.

1.6 Conclusion

The production cost and performance of carbon membranes significantly depend on precursor and carbonization procedures. Carbon membranes made from cellulose precursors present a relatively low cost with moderate separation performance, whereas polyimide-derived carbon membranes in general show high performance, but the production cost might be relatively high. The microstructures of carbon membranes can be easily implemented either during carbonization or by an extra post-treatment process, which makes the carbon membrane more flexible in different separation applica-tions. However, membrane aging caused by the adsorption of gas molecules inside the carbon matrix should be considered for long-term use, and suit-able regeneration methods should be employed to recover the membrane performance over time. The main challenges of carbon membrane upscaling are to establish a low-cost continuous carbonization process and to secure a module design that is suitable for high-pressure and high-temperature applications.

Acknowledgments

The authors would like to thank the Research Council of Norway for fund-ing this work through the PETROMAKS 2 program in the CO2Hing project (#267615).

References

[1] P.H.T. Ngamou, M.E. Ivanova, O. Guillon, W.A. Meulenberg, High-performance carbon molecular sieve membranes for hydrogen purification and pervaporation dehydration of organic solvents, *Journal of Materials Chemistry A*, 7 (2019) 7082–7091.

[2] Z. Yang, W. Guo, S.M. Mahurin, S. Wang, H. Chen, L. Cheng, K. Jie, H.M. Meyer, D.-e. Jiang, G. Liu, W. Jin, I. Popovs, S. Dai, Surpassing Robeson upper limit for CO_2/N_2 separation with fluorinated carbon molecular sieve membranes, *Chem*, 6 (2020) 631–645.

[3] W.J. Koros, C. Zhang, Materials for next-generation molecularly selective synthetic membranes, *Nature Materials*, 16 (2017) 289–297.

[4] S.M. Saufi, A.F. Ismail, Fabrication of carbon membranes for gas separation - A review, *Carbon*, 42 (2004) 241–259.

[5] M.-B. Hagg, J.A. Lie, A. Lindbrathen, Carbon molecular sieve membranes. A promising alternative for selected industrial applications, *Annals of the New York Academy of Sciences*, 984 (2003) 329–345.

[6] J.E. Koresh, A. Soffer, Molecular sieve carbon membrane Part I: Presentation of a new device for gas mixture separation, *Separation Science and Technology*, 18 (1983) 723–734.

[7] J.A. Lie, M.-B. Hagg, Carbon membranes from cellulose: Synthesis, performance and regeneration, *Journal of Membrane Science*, 284 (2006) 79–86.

[8] X. He, J.A. Lie, E. Sheridan, M.-B. Hägg, Preparation and characterization of hollow fiber carbon membranes from cellulose acetate precursors, *Industrial & Engineering Chemistry Research*, 50 (2011) 2080–2087.

[9] X. He, M.-B. Hägg, Structural, kinetic and performance characterization of hollow fiber carbon membranes, *Journal of Membrane Science*, 390–391 (2012) 23–31.

[10] X. He, M.-B. Hagg, Optimization of carbonization process for preparation of high performance hollow fiber carbon membranes, *Industrial & Engineering Chemistry Research*, 50 (2011) 8065–8072.

[11] L. Lei, A. Lindbråthen, M. Hillestad, M. Sandru, E.P. Favvas, X. He, Screening cellulose spinning parameters for fabrication of novel carbon hollow fiber membranes for gas separation, *Industrial & Engineering Chemistry Research*, 58 (2019) 13330–13339.

[12] S.C. Rodrigues, M. Andrade, J. Moffat, F.D. Magalhães, A. Mendes, Preparation of carbon molecular sieve membranes from an optimized ionic liquid-regenerated cellulose precursor, *Journal of Membrane Science*, 572 (2019) 390–400.

[13] M. Rungta, L. Xu, W.J. Koros, Carbon molecular sieve dense film membranes derived from Matrimid® for ethylene/ethane separation, *Carbon*, 50 (2012) 1488–1502.

[14] N. Bhuwania, Y. Labreche, C.S.K. Achoundong, J. Baltazar, S.K. Burgess, S. Karwa, L. Xu, C.L. Henderson, P.J. Williams, W.J. Koros, Engineering substructure morphology of asymmetric carbon molecular sieve hollow fiber membranes, *Carbon*, 76 (2014) 417–434.

[15] Y.-J. Fu, K.-S. Liao, C.-C. Hu, K.-R. Lee, J.-Y. Lai, Development and characterization of micropores in carbon molecular sieve membrane for gas separation, *Microporous and Mesoporous Materials*, 143 (2011) 78–86.

[16] M.G. Kamath, S. Fu, A.K. Itta, W. Qiu, G. Liu, R. Swaidan, W.J. Koros, 6FDA-DETDA: DABE polyimide-derived carbon molecular sieve hollow fiber membranes: Circumventing unusual aging phenomena, *Journal of Membrane Science*, 546 (2018) 197–205.

[17] Y. Cao, K. Zhang, O. Sanyal, W.J. Koros, Carbon molecular sieve membrane preparation by economical coating and pyrolysis of porous polymer hollow fibers, *Angewandte Chemie International Edition*, 58 (2019) 12149–12153.

[18] W. Qiu, J. Vaughn, G. Liu, L. Xu, M. Brayden, M. Martinez, T. Fitzgibbons, G. Wenz, W.J. Koros, Hyperaging Tuning of a carbon molecular-sieve hollow fiber membrane with extraordinary gas-separation performance and stability, *Angewandte Chemie International Edition*, 58 (2019) 11700–11703.

[19] C. Zhang, W.J. Koros, Ultraselective carbon molecular sieve membranes with tailored synergistic sorption selective properties, *Advanced Materials*, 29 (2017) 1701631.

[20] D.-Y. Koh, B.A. McCool, H.W. Deckman, R.P. Lively, Reverse osmosis molecular differentiation of organic liquids using carbon molecular sieve membranes, *Science*, 353 (2016) 804.

[21] L.I.B. David, A.F. Ismail, Influence of the thermastabilization process and soak time during pyrolysis process on the polyacrylonitrile carbon membranes for O_2/N_2 separation, *Journal of Membrane Science*, 213 (2003) 285–291.

[22] M. Rungta, G.B. Wenz, C. Zhang, L. Xu, W. Qiu, J.S. Adams, W.J. Koros, Carbon molecular sieve structure development and membrane performance relationships, *Carbon*, 115 (2017) 237–248.

[23] L. Xu, M. Rungta, M.K. Brayden, M.V. Martinez, B.A. Stears, G.A. Barbay, W.J. Koros, Olefins-selective asymmetric carbon molecular sieve hollow fiber membranes for hybrid membrane-distillation processes for olefin/paraffin separations, *Journal of Membrane Science*, 423–424 (2012) 314–323.

[24] S. Fu, E.S. Sanders, S.S. Kulkarni, W.J. Koros, Carbon molecular sieve membrane structure–property relationships for four novel 6FDA based polyimide precursors, *Journal of Membrane Science*, 487 (2015) 60–73.

[25] X. He, M.-B. Hägg, Hollow fiber carbon membranes: Investigations for CO_2 capture, *Journal of Membrane Science*, 378 (2011) 1–9.

[26] X. He, M.-B. Hägg, Hollow fiber carbon membranes: From material to application, *Chemical Engineering Journal*, 215–216 (2013) 440–448.

[27] J.S. Adams, A.K. Itta, C. Zhang, G.B. Wenz, O. Sanyal, W.J. Koros, New insights into structural evolution in carbon molecular sieve membranes during pyrolysis, *Carbon*, 141 (2019) 238–246.

[28] D.Q. Vu, W.J. Koros, S.J. Miller, High pressure CO_2/CH_4 separation using carbon molecular sieve hollow fiber membranes, *Industrial & Engineering Chemistry Research*, 41 (2002) 367–380.

[29] C. Zhang, G.B. Wenz, P.J. Williams, J.M. Mayne, G. Liu, W.J. Koros, Purification of aggressive supercritical natural gas using carbon molecular sieve hollow fiber membranes, *Industrial & Engineering Chemistry Research*, 56 (2017) 10482–10490.

[30] B. Zhang, Y. Wu, Y. Lu, T. Wang, X. Jian, J. Qiu, Preparation and characterization of carbon and carbon/zeolite membranes from ODPA–ODA type polyetherimide, *Journal of Membrane Science*, 474 (2015) 114–121.

[31] V.M. Linkov, R.D. Sanderson, E.P. Jacobs, Highly asymmetrical carbon membranes, *Journal of Membrane Science*, 95 (1994) 93–99.

[32] H.-J. Lee, H. Suda, K. Haraya, S.-H. Moon, Gas permeation properties of carbon molecular sieving membranes derived from the polymer blend of polyphenylene oxide (PPO)/polyvinylpyrrolidone (PVP), *Journal of Membrane Science*, 296 (2007) 139–146.

[33] M. Yoshimune, I. Fujiwara, K. Haraya, Carbon molecular sieve membranes derived from trimethylsilyl substituted poly(phenylene oxide) for gas separation, *Carbon*, 45 (2007) 553–560.

[34] A.K. Itta, H.-H. Tseng, M.-Y. Wey, Fabrication and characterization of PPO/PVP blend carbon molecular sieve membranes for H_2/N_2 and H_2/CH_4 separation, *Journal of Membrane Science*, 372 (2011) 387–395.

[35] T.A. Centeno, A.B. Fuertes, Supported carbon molecular sieve membranes based on a phenolic resin, *Journal of Membrane Science*, 160 (1999) 201–211.

[36] T.A. Centeno, J.L. Vilas, A.B. Fuertes, Effects of phenolic resin pyrolysis conditions on carbon membrane performance for gas separation, *Journal of Membrane Science*, 228 (2004) 45–54.

[37] H.B. Park, Y.K. Kim, J.M. Lee, S.Y. Lee, Y.M. Lee, Relationship between chemical structure of aromatic polyimides and gas permeation properties of their carbon molecular sieve membranes, *Journal of Membrane Science*, 229 (2004) 117–127.

[38] C.-P. Hu, C.K. Polintan, L.L. Tayo, S.-C. Chou, H.-A. Tsai, W.-S. Hung, C.-C. Hu, K.-R. Lee, J.-Y. Lai, The gas separation performance adjustment of carbon molecular sieve membrane depending on the chain rigidity and free volume characteristic of the polymeric precursor, *Carbon*, 143 (2019) 343–351.

[39] J. Gao, Y. Wang, H. Wu, X. Liu, L. Wang, Q. Yu, A. Li, H. Wang, C. Song, Z. Gao, M. Peng, M. Zhang, N. Ma, J. Wang, W. Zhou, G. Wang, Z. Yin, D. Ma, Construction of a sp^3/sp^2 Carbon interface in 3D N-doped nanocarbons for the oxygen reduction reaction, *Angewandte Chemie International Edition*, 58 (2019) 15089–15097.

[40] L.M. Robeson, The upper bound revisited, *Journal of Membrane Science*, 320 (2008) 390–400.

[41] N. Peng, N. Widjojo, P. Sukitpaneenit, M.M. Teoh, G.G. Lipscomb, T.-S. Chung, J.-Y. Lai, Evolution of polymeric hollow fibers as sustainable technologies: Past, present, and future, *Progress in Polymer Science*, 37 (2012) 1401–1424.

[42] L. Xu, M. Rungta, J. Hessler, W. Qiu, M. Brayden, M. Martinez, G. Barbay, W.J. Koros, Physical aging in carbon molecular sieve membranes, *Carbon*, 80 (2014) 155–166.

[43] Y. Ma, M.L. Jue, F. Zhang, R. Mathias, H.Y. Jang, R.P. Lively, Creation of well-defined "mid-sized" micropores in carbon molecular sieve membranes, *Angewandte Chemie International Edition*, 131 (2019) 13393–13399.

[44] J. Zemek, J. Houdkova, P. Jiricek, M. Jelinek, Surface and in-depth distribution of sp^2 and sp^3 coordinated carbon atoms in diamond-like carbon films modified by argon ion beam bombardment during growth, *Carbon*, 134 (2018) 71–79.

[45] H.-J. Lee, D.-P. Kim, H. Suda, K. Haraya, Gas permeation properties for the post-oxidized polyphenylene oxide (PPO) derived carbon membranes: Effect of the oxidation temperature, *Journal of Membrane Science*, 282 (2006) 82–88.

[46] H. Richter, H. Voss, N. Kaltenborn, S. Kämnitz, A. Wollbrink, A. Feldhoff, J. Caro, S. Roitsch, I. Voigt, High-flux carbon molecular sieve membranes for gas separation, *Angewandte Chemie International Edition*, 56 (2017) 7760–7763.

[47] S. Haider, A. Lindbråthen, J.A. Lie, I.C.T. Andersen, M.-B. Hägg, CO_2 separation with carbon membranes in high pressure and elevated temperature applications, *Separation and Purification Technology*, 190 (2018) 177–189.

[48] A.B. Fuertes, Effect of air oxidation on gas separation properties of adsorption-selective carbon membranes, *Carbon*, 39 (2001) 697–706.

[49] M. Kiyono, P.J. Williams, W.J. Koros, Effect of pyrolysis atmosphere on separation performance of carbon molecular sieve membranes, *Journal of Membrane Science*, 359 (2010) 2–10.

[50] R. Singh, W.J. Koros, Carbon molecular sieve membrane performance tuning by dual temperature secondary oxygen doping (DTSOD), *Journal of Membrane Science*, 427 (2013) 472–478.

[51] S. Huang, L.F. Villalobos, D.J. Babu, G. He, M. Li, A. Züttel, K.V. Agrawal, Ultrathin carbon molecular sieve films and room-temperature oxygen functionalization for gas-sieving, *ACS Applied Materials & Interfaces*, 11 (2019) 16729–16736.

[52] G.B. Wenz, W.J. Koros, Tuning carbon molecular sieves for natural gas separations: A diamine molecular approach, *AIChE Journal*, 63 (2017) 751–760.

[53] K. Hazazi, X. Ma, Y. Wang, W. Ogieglo, A. Alhazmi, Y. Han, I. Pinnau, Ultra-selective carbon molecular sieve membranes for natural gas separations based on a carbon-rich intrinsically microporous polyimide precursor, *Journal of Membrane Science*, 585 (2019) 1–9.

[54] O. Salinas, X. Ma, Y. Wang, Y. Han, I. Pinnau, Carbon molecular sieve membrane from a microporous spirobisindane-based polyimide precursor with enhanced ethylene/ethane mixed-gas selectivity, *RSC Advances*, 7 (2017) 3265–3272.

[55] J.A. Lie, X. He, I. Kumakiri, H. Kita, M.-B. Hagg, Carbon-based membranes, in: *Hydrogen Production, Separation and Purification for Energy*, Institution of Engineering and Technology, 2017, pp. 405–431.

[56] C.W. Jones, W.J. Koros, Carbon molecular sieve gas separation membranes-II. Regeneration following organic exposure, *Carbon*, 32 (1994) 1427–1432.

[57] S. Haider, A. Lindbråthen, J.A. Lie, P.V. Carstensen, T. Johannessen, M.-B. Hägg, Vehicle fuel from biogas with carbon membranes; a comparison between simulation predictions and actual field demonstration, *Green Energy & Environment*, 3 (2018) 266–276.

[58] O. Karvan, J.R. Johnson, P.J. Williams, W.J. Koros, A pilot-scale system for carbon molecular sieve hollow fiber membrane manufacturing, *Chemical Engineering and Technology*, 36 (2013) 53–61.

[59] S. Haider, J.A. Lie, A. Lindbråthen, M.-B. Hägg, Pilot–scale production of carbon hollow fiber membranes from regenerated cellulose precursor-Part II: Carbonization procedure, *Membranes*, 8 (2018) 97.

[60] D. Parsley, R.J. Ciora Jr, D.L. Flowers, J. Laukaitaus, A. Chen, P.K.T. Liu, J. Yu, M. Sahimi, A. Bonsu, T.T. Tsotsis, Field evaluation of carbon molecular sieve

membranes for the separation and purification of hydrogen from coal- and bio-mass-derived syngas, *Journal of Membrane Science*, 450 (2014) 81–92.

[61] Abraham Soffer, Jack Gilron, Shlomo Saguee, Rafael Hed-Ofek, H. Cohen, *Process for the production of hollow carbon fiber membranes*, US5925591A (1995).

[62] S. Haider, A. Lindbråthen, J.A. Lie, M.-B. Hägg, Carbon membranes for oxygen enriched air – Part I: Synthesis, performance and preventive regeneration, *Separation and Purification Technology*, 204 (2018) 290–297.

[63] S. Haider, A. Lindbråthen, J.A. Lie, M.-B. Hägg, Carbon membranes for oxygen enriched air – Part II: Techno-economic analysis, *Separation and Purification Technology*, 205 (2018) 251–262.

[64] L.I.B. David, A.F. Ismail, Influence of the thermastabilization process and soak time during pyrolysis process on the polyacrylonitrile carbon membranes for O_2/N_2 separation, *Journal of Membrane Science*, 213 (2003) 285–291.

[65] V.M. Linkov, R.D. Sanderson, E.P. Jacobs, Carbon membranes from precursors containing low-carbon residual polymers, *Polymer International*, 35 (1994) 239–242.

[66] H.-H. Tseng, C.-T. Wang, G.-L. Zhuang, P. Uchytil, J. Reznickova, K. Setnickova, Enhanced H_2/CH_4 and H_2/CO_2 separation by carbon molecular sieve membrane coated on titania modified alumina support: Effects of TiO_2 intermediate layer preparation variables on interfacial adhesion, *Journal of Membrane Science*, 510 (2016) 391–404.

[67] M.-Y. Wey, C.-T. Wang, Y.-T. Lin, M.-D. Lin, P. Uchytil, K. Setnickova, H.-H. Tseng, Interfacial interaction between CMS layer and substrate: Critical factors affecting membrane microstructure and H_2 and CO_2 separation performance from CH4, *Journal of Membrane Science*, 580 (2019) 49–61.

[68] C. Zhang, R. Kumar, W.J. Koros, Ultra-thin skin carbon hollow fiber membranes for sustainable molecular separations, *AIChE Journal*, 0 (2019) e16611.

[69] R. Kumar, W.J. Koros, 110th Anniversary: High performance carbon molecular sieve membrane resistance to aggressive feed stream contaminants, *Industrial & Engineering Chemistry Research*, 58 (2019) 6740–6746.

[70] L. Xu, M. Rungta, W.J. Koros, Matrimid® derived carbon molecular sieve hollow fiber membranes for ethylene/ethane separation, *Journal of Membrane Science*, 380 (2011) 138–147.

[71] M. Rungta, L. Xu, W.J. Koros, Structure–performance characterization for carbon molecular sieve membranes using molecular scale gas probes, *Carbon*, 85 (2015) 429–442.

[72] X. Ning, W.J. Koros, Carbon molecular sieve membranes derived from Matrimid® polyimide for nitrogen/methane separation, *Carbon*, 66 (2014) 511–522.

[73] A. Singh-Ghosal, W.J. Koros, Air separation properties of flat sheet homogeneous pyrolytic carbon membranes, *Journal of Membrane Science*, 174 (2000) 177–188.

[74] Y. Sakata, A. Muto, M.A. Uddin, H. Suga, Preparation of porous carbon membrane plates for pervaporation separation applications, *Separation and Purification Technology*, 17 (1999) 97–100.

[75] A.B. Fuertes, Adsorption-selective carbon membrane for gas separation, *Journal of Membrane Science*, 177 (2000) 9–16.

[76] I. Menendez, A.B. Fuertes, Aging of carbon membranes under different environments, *Carbon*, 39 (2001) 733–740.

[77] F.K. Katsaros, T.A. Steriotis, A.K. Stubos, A. Mitropoulos, N.K. Kanellopoulos, S. Tennison, High pressure gas permeability of microporous carbon membranes, *Microporous Materials*, 8 (1997) 171–176.

[78] P.-S. Lee, D. Kim, S.-E. Nam, R.R. Bhave, Carbon molecular sieve membranes on porous composite tubular supports for high performance gas separations, *Microporous and Mesoporous Materials*, 224 (2016) 332–338.

2

Carbon Membrane Performance: State-of-the-Art

Linfeng Lei[a] and Xuezhong He[a,b]

[a]Department of Chemical Engineering, Norwegian University of Science and Technology

[b]Department of Chemical Engineering, Guangdong Technion Israel Institute of Technology (GTIIT)

2.1 Introduction

Carbon membranes with ultra-microporous structures are usually prepared by carbonization of polymeric precursors such as polyimides [1–4] and cellulose derivatives [5–8]. Carbon molecular sieve (CMS) membranes made from polyimide precursors have a potential application in the ultrafine discrimination of molecules that are close in size, due to their high selectivity [3]. However, the production costs of carbon membranes are quite high because of the high cost of polyimide materials. He et al. [5, 9] reported that cellulose acetate (CA)-derived carbon membranes showed good performance in CO_2/CH_4 and CO_2/N_2 separation processes, and their membranes were upscaled to a pilot-scale system with an annual production capacity of 700 m^2 [10]. However, the difficulties in controlling the drying process for the deacetylated hollow fibers remain a challenge for large-scale production.

Spinning cellulose hollow fibers directly from a cellulose dope solution could be a solution. Recently, novel carbon hollow fiber membranes (CHFMs) have been prepared from cellulose precursors directly spun with a cellulose/ (1-ethyl-3-methylimidazolium acetate (EmimAc) + dimethyl sulfoxide) system. The CHFMs show good performances that are above the Robeson upper bound of CO_2/CH_4 and O_2/N_2. The optimized carbon membrane presents a CO_2 permeability of 239 barrer and a CO_2/CH_4 selectivity of 186, based on a pure-gas permeation measurement. Thus, development of cellulose-based CHFMs shows potential for gas separation.

Carbon membranes can be divided into two categories: unsupported and supported. Unsupported membranes have three different configurations: flat-sheet film, hollow fiber, and capillary tubes. Supported carbon membranes have two configurations: flat-sheet and tubular. The unsupported CHFMs are prepared from hollow fiber precursors, which, due to high packing density, could be the only viable configuration when large areas are needed in industrial applications.

2.2 Unsupported Flat-Sheet Carbon Membranes

Unsupported flat-sheet carbon membranes are widely used when investigating the materials' properties because of their relatively easy fabrication process. Solution casting is commonly used to prepare unsupported flat-sheet membranes for laboratory characterization experiments, as illustrated in Figure 2.1. Solution-cast film is produced on a larger scale, as described in a book by Baker [11]. After casting and removing the solvent, the precursors are carbonized to make carbon membranes.

Ning and Koros [2] investigated the sorption and diffusion coefficients of dense CMS films for N_2/CH_4 separation. The dense CMS film carbonized at 800 °C showed an attractive performance that surpassed the Robeson upper bound [12], with N_2 permeability of 6.8 barrer and N_2/CH_4 selectivity of 7.7, based on a pure-gas permeation measurement. Like in other works [3, 5], it was found that gas permeability was reduced and selectivity improved by increasing the carbonization temperature. In addition, gas permeabilities and diffusion coefficients were enhanced with increased testing temperature, following an Arrhenius relationship. However, sorption coefficients were inhibited by increased testing temperature, following a van't Hoff relationship [2]. Moreover, the high selectivity reported in that work showed that the rigid structure of the CMS can provide precise molecular sieving to effectively separate some challenging gas pairs such as N_2/CH_4.

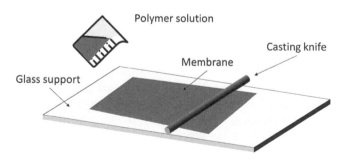

FIGURE 2.1
A schematic showing solution casting for making a flat-sheet film.

Recently, an ultra-selective dense CMS film for natural gas separation based on a carbon-rich intrinsically microporous polyimide precursor (SBFDA-DMN) was reported by Hazazi et al. [13]. Compared with CMS membranes made from a conventional low-free-volume Matrimid polyimide precursor [3], carbon membranes prepared from carbon-rich precursors (with 28 % total weight loss) presented an ultrahigh permeability (>4000 barrer) that is much higher than that of Matrimid (ca. 20 barrer) [13]. However, the two CMS membranes show similar He/CH_4 and CO_2/CH_4 separation performances when carbonized at 800 °C, as is shown in Figure 2.2. The effect of precursor type on separation performance is significant when the carbonization temperature is 900 °C, when the CO_2 permeability for CMS membranes developed from SBFDA-DMN and Matrimid decreased to about 60 % and 42 %, respectively. The authors suggested that the lesser reduction of gas permeability implies that the higher thermal resistance and potentially higher rigidity of the SBFDA-DMN precursor could provide stronger resistance to pore collapse than the Matrimid precursor when carbonized at higher temperatures [13].

Rodrigues et al. [14] reported a dense CMS film that surpasses the Robeson upper bound separation performance for O_2/N_2, He/N_2, and H_2/N_2 and that is derived from an EmimAc-regenerated cellulose precursor. Owing to a highly hydrophilic characteristic that allows water vapor to permeate quickly, the prepared CMS membrane exhibited good humidity stability in the presence of 75–77 % relative humidity.

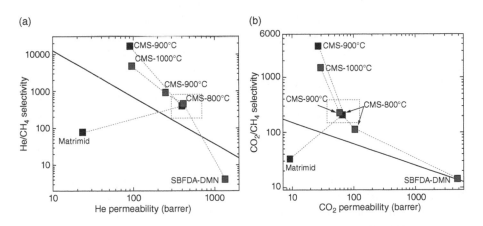

FIGURE 2.2
2008 Robeson upper bounds for (a) He/CH_4 and (b) CO_2/CH_4 separation performances of SBFDA-DMN and Matrimid with their heat-treated CMS samples. The light blue rectangle marks an area where the CMS performance for both precursors is very similar [13].

Although the unsupported flat-sheet CMS membranes exhibit promising gas separation performances, their brittle and fragile characteristics have limited their potential applications on a larger scale, especially compared with supported CMS membranes and hollow fiber configurations.

2.3 Supported Flat-Sheet and Tubular Carbon Membranes

In contrast to unsupported CMS film, which is fragile, supported CMS membranes manufactured along a porous support (e.g., ceramics or metal) exhibit significantly improved mechanical strength. Different options such as dip-coating [15, 16], spin-coating [17–20], spray-coating [21], and chemical vapor deposition (CVD) [22] have been employed to coat the support with polymeric precursor. These methods can reduce the thickness of the membrane selective layer, which usually provides a higher gas permeance than a self-standing film. Representative supported carbon membranes have been reviewed by Lie et al. [23], and the most recently developed membranes are included in Table 2.1.

2.3.1 Support Modification

The properties of porous supports can directly affect the structure of deposited CMS membranes and, consequently, membrane separation performance. For instance, commercial ceramic porous supports are commonly macroporous (pore size ~200 nm), which may lead to interfacial defects when forming the selective layer [24]. Thus, multilayer substrates are widely used for CMS membrane fabrication to improve the interface. A porous intermediate layer (normally 1–10 nm pore size) between the macroporous support and the CMS selective layer is often incorporated during the membrane development process. For example, to avoid large pores in an α-Al_2O_3 tube support that might cause pinholes in the CMS membrane, it is possible to deposit a thin γ-Al_2O_3 layer with mean pore size of ~4 nm [15], which is generally formed by multi dip-coating of a boehmite (γ-AlOOH) sol and followed with a calcination procedure [15, 25, 26].

The separation performance of CMS membranes can be improved by tuning the properties of the intermediate layer. Tseng et al. [27] reported enhanced H_2/CO_2 separation performance by modifying the Al_2O_3 support with a TiO_2 intermediate layer. They suggested that the intermediate layer could provide a networking interlocking pattern with CMS membranes, which is beneficial

TABLE 2.1

Examples of supported CMS membrane preparation conditions and gas permeation results.

Supported support/geometry	Pore size (μm)	Precursor	Method*	Pyrolysis	Gas test temperature	Permeance (10^{-9} mol m^{-2} s^{-1} Pa^{-1})**			Ideal selectivity (separation factor)			Reference
						H_2 (He)	CO_2	O_2	H_2/N_2	O_2/N_2	CO_2/CH_4	
Carbon/disk Φ 35 mm		Matrimid	S	700 °C – 2 h (vacuum)	25 °C	(27)	36.6	8.1	–	4?	8.1 (23)	[42]
Carbon/disk Φ 35 mm		Phenolic resin	S	700 °C (vacuum)	25 °C	(8.2)	2.0	1.2	265		87 (165)	[43]
α-Al₂O₃/tube (o.d. 2.3 mm)	0.14	Phenolic resin + sulfonated phenolic resin	D	500 °C – 1 h (N₂)	35 °C	55 / 134	10 / 40	2.3 / 10	160 / –	10.8 / 12	170 / 54	[44]
α-Al₂O₃/tube (o.d. 10 mm, i.d. 7 mm)	0.2	Novolac resin + bohemite	D	550 °C – 2h (N₂)	Room temperature†	145 (79.5)	–	3.0	725	15	–	[45]
α-Al₂O₃/Tube (o.d. 0.9 mm, i.d. 0.6 mm)	0.005 / 0.0035	Furfuryl alcohol	V	600 °C – 1 h / 600 °C – 1 h	25 °C / 25 °C	25.5 / 6.04	5.82 / 2.67	0.775 / 0.845	347 / 91	10.6 / 12.7	92 / 82	[46]
α-Al₂O₃/o.d. 2.25 mm, i.d. 1.8 mm		Lignocresol	D	600 °C – 1 h (N₂)	35 °C / 105 °C	56 (29) / 82 (43)	17 / 2.3	2.7 / 8.2	167 / 44	8.0 / 4.5	87 / 17	[47]
Anodic alumina/ 4 cm²	0.020	Graphene oxide	V	–	20 °C	100	0.03?	–	~900	–	–	[48]
Coal disk/Φ 40 mm 2 mm thick	0.71 (largest size)	PMDA-ODA	S	800 °C – 2 h		54.55	8.80	7.45	76.3	10.4	–	[49]
α-Al₂O₃ / o.d. 13 mm i.d. 8 mm	3.0	Polyfurfuryl alcohol	D	700 °C – 4 h (Ar)		35	10	7.5	58	12.5	–	[50]

* D: dip-coating, S: spin-coating, V: vapor phase deposition.
** Values were read from figures.
† After air exposure.
†† After H_2O activation.

to both gas permeability and selectivity. A proposed interlocking network for the CMS layer and support is illustrated in Figure 2.3. When the polymeric precursor was deposited on the Al_2O_3 support, the polymer chain penetrated the porous support and formed a dense phase on the surface, as shown in Figure 2.3a. However, the induced intermediate TiO_2 layer could provide interconnected channels with the polymer solution, which reduces the connected depth compared with the Al_2O_3 support, resulting in a thinner and larger d-spaced CMS membrane layer, as shown in Figure 2.3b. Similarly, Wey et al. [20] investigated the interface between the CMS layer and a TiO_2/Al_2O_3 composite support. Owing to the mechanical interlocking effects of TiO_2, the adhesion between the CMS layer and the Al_2O_3 support is improved and thus permeability and selectivity are increased.

2.3.2 Ultrathin Supported CMS Membranes

Developing ultrathin (i.e., <500 nm) CMS selective supported membranes is a promising way to make high-flux CMS membranes. Recently, Ngamou et al. [15] produced an ultrathin (~200 nm) and defect-free CMS membrane by carbonizing a polyimide precursor that was dip-coated on the inner surface of porous supports (γ-Al_2O_3 layer coating α-Al_2O_3 tubes). The prepared CMS membrane showed a high H_2 permeance of up to 3253 GPU and ideal selectivities of 24, 130, and 228 for H_2/CO_2, H_2/N_2, and H_2/CH_4, respectively, at 200 °C. Furthermore, the supported CMS exhibited good separation performance for pervaporation when applied to a 10 wt % water-containing

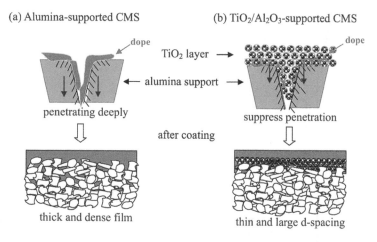

FIGURE 2.3
The proposed interlocking pattern for a CMS membrane supported on (a) alumina and (b) TiO_2/Al_2O_3 supports [27].

binary mixture of methanol and ethanol. Separation factors of 791 and 1946 for methanol and ethanol, respectively, were obtained at 70 °C with a water flux of about 0.5 kg m^{-2} h^{-1}.

A 100-nm-thick CMS film was produced by Huang et al. [28] who employed a novel fabrication route, named the "transfer technique," leading to an attractive H$_2$ permeance of 3060 GPU. The fabrication approach is depicted in Figure 2.4. The polymeric precursor film, which was spin-coated on a Cu foil support, was carbonized at 500 °C in a H$_2$/Ar atmosphere to obtain a CMS film. The Cu foil was subsequently removed by etching in 1 M FeCl3 solution and the free-standing CMS film rinsed with deionized water. The CMS film was washed with deionized water and supported on a porous W support for testing. The separation performance was further improved by an ozone post treatment that shrank the electron density gap in the nanopores. The optimized membranes yielded a H$_2$ permeance of 507 GPU with a H$_2$/CO$_2$ selectivity of 7.1.

Hou et al. [18] recently reported the fabrication of CMS membranes with a 100-nm-thick layer. In their work, carbon nanotubes (CNTs) were employed as efficient scaffolds to prevent the polymer precursor

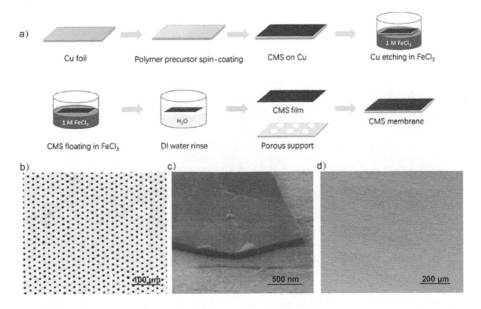

FIGURE 2.4
Ultrathin CMS film using the transfer approach. (a) Schematic of the fabrication of ultrathin CMS membranes by the transfer method. (b) Scanning electron microscope (SEM) image of a macroporous W support. (c) Cross-sectional SEM image of a CMS film prepared in this fashion. (d) Top-view SEM image of the CMS membranes on a macroporous W support [28].

(poly(furfuryl alcohol), PFA) penetrating the pores of the anodized aluminum oxide (AAO) support, as illustrated in Figure 2.5. The dispersed CNT solution was vacuum-filtered on the AAO support; the CMS/CNT hybrid membrane was then fabricated by carbonization of a PFA precursor that was spin-coated. As shown in Figure 2.5a–c, with the help of CNTs, a continuous CMS/CNT layer was obtained with no carbon deposited in the AAO channels. Conversely, most PFA solution penetrated the AAO channels after being spread on the bare AAO support, resulting in irregular blocks and defects and not a continuous CMS selective layer. Moreover, by controlling CNT loading, the thickness of the CMS/CNT hybrid active layer can be changed from 100 to 1000 nm. An optimized membrane with a thickness of 322 nm exhibited O_2 permeance of 135 GPU and O_2/N_2 selectivity of 10.5 [18].

Overall, although the fabrication of defect-free ultrathin CMS membranes is still challenging, as the thin layer is fragile, the development of supported CMS membranes with an ultrathin selective layer shows a very promising approach to improve gas permeance. However, thinner CMS membranes normally lead to lower selectivity, which is not conducive to separation efficiency. To improve separation selectivity, in addition to exploiting new polymer precursors, modifying the porous support (such as enhancing chemical bonding interaction between the support and CMS layer) and post treatment for CMS membrane preparation may be employed.

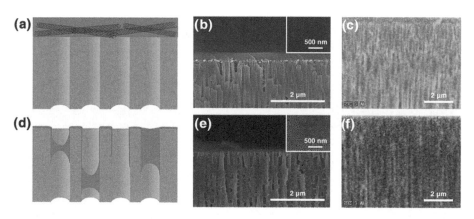

FIGURE 2.5
Key role of the CNT network in the preparation of an ultrathin, continuous CMS membrane. (a, d) Schematic illustration of the spreading of PFA on the AAO substrate (a) with or (d) without the CNT network. The blue regions denote PFA. (b) Cross-sectional and top-view (inset) SEM images and (c) Energy-dispersive X-ray spectroscopy mapping image of the CMS/CNT hybrid membrane with a continuous layer on the AAO substrate. (e) Cross-sectional and top-view (inset) SEM images and (f) EDX mapping image of the carbonized PFA-coated AAO without the CNT network [18].

2.4 Carbon Hollow Fiber Membranes

Membranes with the hollow fiber configuration have the advantage of packing density compared with spiral wound, plate-and-frame, and tubular modules, with 30,000 m^2 m^{-3} configurations possible [29], which is the industrially preferred configuration. Carbon membranes in the form of hollow fibers have the ability to withstand high transmembrane pressures [30]. Thus, CHFMs show promising applications in different separation scenarios, such as CO_2 removal from natural gas [31–34], organic solvent reverse osmosis [35], and H_2 purification [36, 37]. Hollow fiber precursor spinning and the following carbonization process are essential to fabricate high-performance CHFMs in symmetric or asymmetric structures. Normally, to obtain symmetric CHFMs, anti-collapse treatment during the carbonization process is needed.

2.4.1 Symmetric CHFMs

Symmetric CHFMs with a thicker selective layer normally provide remarkable gas-pair selectivity compared with asymmetric structures with a skin selective layer. Zhang et al. [3] reported ultrahigh permselectivities in CHFMs derived from Matrimid polyimide precursors. The reported CHFMs display a symmetric structure with a well-defined separation layer and properties, which are convenient for understanding the material's intrinsic permeation properties. The membrane carbonized at 900 °C presents the ideal selectivity for different gas pairs: α [CO_2/CH_4] = 3650, α [N_2/CH_4] = 63, α [O_2/N_2] = 21, α [He/CH_4] = 16700, α [H_2/CH_4] = 40350, and α [H_2/N_2] = 640, which are the highest selectivities reported to date. Using the time-lag method, the authors deconvoluted gas permeability into diffusivity and sorption coefficients to explain such unprecedentedly high selectivities. Based on the calculated diffusivities and sorption coefficients, the selectivities of different gas pairs were also divided into diffusion selectivity and sorption selectivity. The authors found that that increased CO_2/CH_4 selectivity following increased carbonization temperature is due to simultaneously improved diffusion selectivity and sorption selectivity. On the one hand, elevating carbonization temperature tends to form a more ordered graphitic carbon structure (sp^2 hybridized carbon) that results in denser packing and smaller micropore and ultramicropore sizes. The refined ultramicropores improve the gas-pair diffusion selectivity. On the other hand, the smaller pore structure restricts sorption of larger gases like CH_4. As a result, both diffusion selectivity and sorption selectivity are enhanced.

2.4.2 Asymmetric CHFMs

Asymmetric CHFMs are attractive to achieve high gas permeance with decent selectivity as well as maintain mechanical strength. Recently, Zhang et al. [31] reported a polyimide-derived CHFM with CO_2/CH_4 separation factors ~60 under a supercritical (1800 psia) natural gas feed comprising 50% CO_2 and 500 ppm highly condensable C7 hydrocarbons. With attractive and stable separation performance, the CHFM can potentially enable CO_2 removal processes in challenging natural gas feeds.

The asymmetric structure of a precursor can be developed by adjusting the spinning conditions. In order to maintain the asymmetric structure during the carbonization process, an extra pretreatment step before carbonization, referred to as "crosslinking," is usually required. Koh et al. [35] reported asymmetric CHFMs made from polyvinylidene difluoride (PVDF) hollow fibers. The non-crosslinked porous PVDF precursors were found to form highly dense hollow fiber membranes with an irreversible loss of porosity during carbonization. They suggested that the porous structure collapsing in the inside membrane was caused by the loss of PVDF storage modulus when heating to 100 °C, which is lower than the melting point (ca. 165 °C) [35]. To overcome such a challenge, a crosslinking method was used to form covalent bonds between PVDF chains by introducing a base treatment (NaOH + MeOH) and following nucleophilic attack with para-xylylenediamine. Compared with the neat PVDF hollow fibers, they reported that the crosslinked PVDF precursors maintained a high storage modulus when heated above 300 °C, and the asymmetric porous structure of the hollow fiber membranes was retained after carbonization.

The Koros group [34, 38–40] used a V-treatment method to restrict the morphology shrinking of asymmetric hollow fiber membranes during the carbonization process. This method avoids a chemical reaction between the polymer precursor and the agent because the sol–gel crosslinking reaction takes place between the organic-alkoxy silane (vinyltrimethoxysilane) and moisture at room temperature, forming a vinyl-crosslinked silica layer. The crosslinked layer provides reinforced sheaths on the "struts," and thus restricts substructure collapse during carbonization. The thickness of the selective layer of asymmetric hollow fiber membrane has been reduced by 5–6 times by the V-treatment. The gas permeance of asymmetric CHFMs was four times better compared with the non-treated membranes in permeation testing, both with single gases and with aggressive high-pressure mixed CO_2/CH_4, while the CO_2/CH_4 selectivity slightly decreased [38].

However, the associated crosslinking steps may account for a cost increment of ~40% to the overall membrane fabrication process [41]. Thus, the development of CHFMs without a need for extra treatment could significantly improve the yield of membranes and reduce cost. Recently, Jue et al. [41] reported asymmetric CHFMs derived directly from PIM-1 precursors, without crosslinking or other pretreatments. They suggested that the similarity of the glass transition temperature and the decomposition temperature of the polymer can maintain the asymmetric structure during the carbonization process, as carbonization happens before the significant changes of storage modulus [41].

Most CHFMs reported so far still present a relatively thick selective layer (>3 μm), which restricts the gas permeance. Further reducing the thickness of CHFMs can provide significantly enhanced gas permeances and thus reduce the required membrane area for a specific application. In order to reduce the thickness of the skin layer, Kumar et al. reported a dual-layer precursor spinning process by coextruding a sheath polymer dope and a core polymer dope from a multichannel spinneret [39]. The dual-layer structure of the precursor was well maintained after V-treatment and carbonization. As a result, CHFMs with ultrathin skin layers (~500 nm) could be created.

An even thinner sheath layer can be obtained by reducing the flow rate ratio of sheath to core dope. Additionally, to reduce the cost of polymer materials, the authors proposed dual-layer precursor spinning, which comprises different sheath and core layer polymers. Considering the core layer as a porous support for the skin layer, the relatively inexpensive Matrimid polyimide was used only in the core dope, while the sheath dope comprises the 6FDA/BPDA-DAM polyimide (4,4-(hexafluoroisopropylidene) diphthalic anhydride (6FDA)–3,3-4,4-biphenyl tetracarboxylic acid dianhydride (BPDA)–2,4,6-trimethyl-1,3-phenylene diamine (DAM)). Cao et al. [34] reported a composite precursor that is composed of a high-cost dense skin layer created by dip-coating a cheaper porous support fabricated by spinning, which can reduce the material cost 25-fold compared with a monolithic precursor. The composite precursor provides a carbon membrane with a 300 nm selective layer, which exhibits excellent separation performance.

2.5 Conclusion

Both self-supported and ceramic-supported carbon membranes in flat-sheet and hollow fiber configurations are found to be promising for different separations, such as CO_2/CH_4, CO_2/N_2, H_2/CO_2, and alkene/alkane separations. However, none of the developed carbon membranes have been successfully brought to the market on an industrial scale either because of the high production cost or the complex fabrication process. Although the most promising precursors of polyimide and cellulose identified balance the production cost and separation performance, further work is still needed to reduce production costs and reduce the number of process steps. Methods to develop asymmetric CHFMs that avoid complex pretreatments to maintain their asymmetric structure may have great potential to overcome the issues. Finally, it is suggested that self-supported CHFMs may have lower costs and applications for larger gas volumes, whereas supported carbon membranes will probably be a good choice for small gas volumes due to their lower packing density.

Acknowledgments

The authors would like to thank the Research Council of Norway for funding this work through the PETROMAKS 2 program in the CO2Hing project (#267615).

References

[1] E.P. Favvas, G.E. Romanos, S.K. Papageorgiou, F.K. Katsaros, A.C. Mitropoulos, N.K. Kanellopoulos, A methodology for the morphological and physicochemical characterisation of asymmetric carbon hollow fiber membranes, *Journal of Membrane Science*, 375 (2011) 113–123.

[2] X. Ning, W.J. Koros, Carbon molecular sieve membranes derived from Matrimid® polyimide for nitrogen/methane separation, *Carbon*, 66 (2014) 511–522.

[3] C. Zhang, W.J. Koros, Ultraselective carbon molecular sieve membranes with tailored synergistic sorption selective properties, *Advanced Materials*, 29 (2017) 1701631.

[4] J.S. Adams, A.K. Itta, C. Zhang, G.B. Wenz, O. Sanyal, W.J. Koros, New insights into structural evolution in carbon molecular sieve membranes during pyrolysis, *Carbon*, 141 (2019) 238–246.

[5] X. He, M.-B. Hägg, Optimization of carbonization process for preparation of high performance hollow fiber carbon membranes, *Industrial & Engineering Chemistry Research* 50 (2011) 8065–8072.

[6] X. He, J.A. Lie, E. Sheridan, M.-B. Hägg, Preparation and characterization of hollow fiber carbon membranes from cellulose acetate precursors, *Industrial & Engineering Chemistry Research*, 50 (2011) 2080–2087.

[7] X. He, M.-B. Hägg, Hollow fiber carbon membranes: Investigations for CO_2 capture, *Journal of Membrane Science*, 378 (2011) 1–9.

[8] X. He, T.-J. Kim, M.-B. Hägg, Hybrid fixed-site-carrier membranes for CO_2 removal from high pressure natural gas: Membrane optimization and process condition investigation, *Journal of Membrane Science*, 470 (2014) 266–274.

[9] X. He, M.-B. Hägg, Hollow fiber carbon membranes: From material to application, *Chemical Engineering Journal*, 215–216 (2013) 440–448.

[10] S. Haider, J. Lie, A. Lindbråthen, M.-B.J.M. Hägg, Pilot–scale production of carbon hollow fiber membranes from regenerated cellulose precursor-Part I: Optimal conditions for precursor preparation, *Membranes*, 8 (2018) 105.

[11] R.W. Baker, Overview of membrane science and technology, in: *Membrane Technology and Applications*, 2004, pp. 1–14.

[12] L.M. Robeson, The upper bound revisited, *Journal of Membrane Science*, 320 (2008) 390–400.

[13] K. Hazazi, X. Ma, Y. Wang, W. Ogieglo, A. Alhazmi, Y. Han, I. Pinnau, Ultra-selective carbon molecular sieve membranes for natural gas separations based on a carbon-rich intrinsically microporous polyimide precursor, *Journal of Membrane Science*, 585 (2019) 1–9.

[14] S.C. Rodrigues, M. Andrade, J. Moffat, F.D. Magalhães, A. Mendes, Preparation of carbon molecular sieve membranes from an optimized ionic liquid-regenerated cellulose precursor, *Journal of Membrane Science*, 572 (2019) 390–400.

[15] P.H.T. Ngamou, M.E. Ivanova, O. Guillon, W.A. Meulenberg, High-performance carbon molecular sieve membranes for hydrogen purification and pervaporation dehydration of organic solvents, *Journal of Materials Chemistry A*, 7 (2019) 7082–7091.

[16] S. Tanaka, T. Yasuda, Y. Katayama, Y. Miyake, Pervaporation dehydration performance of microporous carbon membranes prepared from resorcinol/formaldehyde polymer, *Journal of Membrane Science*, 379 (2011) 52–59.

[17] K. Briceño, D. Montané, R. Garcia-Valls, A. Iulianelli, A. Basile, Fabrication variables affecting the structure and properties of supported carbon molecular sieve membranes for hydrogen separation, *Journal of Membrane Science*, 415–416 (2012) 288–297.

[18] J. Hou, H. Zhang, Y. Hu, X. Li, X. Chen, S. Kim, Y. Wang, G.P. Simon, H. Wang, Carbon nanotube networks as nanoscaffolds for fabricating ultrathin carbon molecular sieve membranes, *ACS Applied Materials & Interfaces*, 10 (2018) 20182–20188.

[19] W. Ogieglo, A. Furchner, X. Ma, K. Hazazi, A.T. Alhazmi, I. Pinnau, Thin composite carbon molecular sieve membranes from a polymer of intrinsic microporosity precursor, *ACS Applied Materials & Interfaces*, 11 (2019) 18770–18781.

[20] M.-Y. Wey, C.-T. Wang, Y.-T. Lin, M.-D. Lin, P. Uchytil, K. Setnickova, H.-H. Tseng, Interfacial interaction between CMS layer and substrate: Critical factors affecting membrane microstructure and H_2 and CO_2 separation performance from CH4, *Journal of Membrane Science*, 580 (2019) 49–61.

[21] N.H. Ismail, W.N.W. Salleh, N. Sazali, A.F. Ismail, N. Yusof, F. Aziz, Disk supported carbon membrane via spray coating method: Effect of carbonization temperature and atmosphere, *Separation and Purification Technology*, 195 (2018) 295–304.

[22] Y. Xie, L. Lu, Y. Tang, Q. Chen, Grafting carbon nanotubes onto copper fibers using a one-step chemical vapor deposition process, *Materials Letters*, 153 (2015) 96–98.

[23] J.A. Lie, X. He, I. Kumakiri, H. Kita, M.-B. Hagg, Carbon-based membranes, in: *Hydrogen Production, Separation and Purification for Energy*, Institution of Engineering and Technology, 2017, pp. 405–431.

[24] J.B.S. Hamm, A. Ambrosi, J.G. Griebeler, N.R. Marcilio, I.C. Tessaro, L.D. Pollo, Recent advances in the development of supported carbon membranes for gas separation, *International Journal of Hydrogen Energy*, 42 (2017) 24830–24845.

[25] B.S. Liu, N. Wang, F. He, J.X. Chu, Separation performance of nanoporous carbon membranes fabricated by catalytic decomposition of CH_4 using Ni/polyamideimide templates, *Industrial & Engineering Chemistry Research*, 47 (2008) 1896–1902.

[26] X. Ma, B.K. Lin, X. Wei, J. Kniep, Y.S. Lin, Gamma-alumina supported carbon molecular sieve membrane for propylene/propane separation, *Industrial & Engineering Chemistry Research*, 52 (2013) 4297–4305.

[27] H.-H. Tseng, C.-T. Wang, G.-L. Zhuang, P. Uchytil, J. Reznickova, K. Setnickova, Enhanced H_2/CH_4 and H_2/CO_2 separation by carbon molecular sieve membrane coated on titania modified alumina support: Effects of TiO_2 intermediate layer preparation variables on interfacial adhesion, *Journal of Membrane Science*, 510 (2016) 391–404.

[28] S. Huang, L.F. Villalobos, D.J. Babu, G. He, M. Li, A. Züttel, K.V. Agrawal, Ultrathin carbon molecular sieve films and room-temperature oxygen functionalization for gas-sieving, *ACS Applied Materials & Interfaces*, 11 (2019) 16729–16736.

[29] M. Mulder, Module and process design, in: M. Mulder (Ed.) *Basic Principles of Membrane Technology*, Springer, Netherlands, Dordrecht, 1996, pp. 465–520.

[30] Y. Ma, M.L. Jue, F. Zhang, R. Mathias, H.Y. Jang, R.P. Lively, Creation of well-defined "mid-sized" micropores in carbon molecular sieve membranes, *Angewandte Chemie International Edition*, 131 (2019) 13393–13399.

[31] C. Zhang, G.B. Wenz, P.J. Williams, J.M. Mayne, G. Liu, W.J. Koros, Purification of aggressive supercritical natural gas using carbon molecular sieve hollow fiber membranes, *Industrial & Engineering Chemistry Research*, 56 (2017) 10482–10490.

[32] L. Lei, A. Lindbråthen, M. Hillestad, M. Sandru, E.P. Favvas, X. He, Screening cellulose spinning parameters for fabrication of novel carbon hollow fiber membranes for gas separation, *Industrial & Engineering Chemistry Research*, 58 (2019) 13330–13339.

[33] G.B. Wenz, W.J. Koros, Tuning carbon molecular sieves for natural gas separations: A diamine molecular approach, *AIChE Journal*, 63 (2017) 751–760.

[34] Y. Cao, K. Zhang, O. Sanyal, W.J. Koros, Carbon molecular sieve membrane preparation by economical coating and pyrolysis of porous polymer hollow fibers, *Angewandte Chemie International Edition*, 58 (2019) 12149–12153.

[35] D.-Y. Koh, B.A. McCool, H.W. Deckman, R.P. Lively, Reverse osmosis molecular differentiation of organic liquids using carbon molecular sieve membranes, *Science*, 353 (2016) 804.

[36] E.P. Favvas, N.S. Heliopoulos, S.K. Papageorgiou, A.C. Mitropoulos, G.C. Kapantaidakis, N.K. Kanellopoulos, Helium and hydrogen selective carbon hollow fiber membranes: The effect of pyrolysis isothermal time, *Separation and Purification Technology*, 142 (2015) 176–181.

[37] S. Sá, J.M. Sousa, A. Mendes, Steam reforming of methanol over a CuO/ZnO/Al_2O_3 catalyst part II: A carbon membrane reactor, *Chemical Engineering Science*, 66 (2011) 5523–5530.

[38] N. Bhuwania, Y. Labreche, C.S.K. Achoundong, J. Baltazar, S.K. Burgess, S. Karwa, L. Xu, C.L. Henderson, P.J. Williams, W.J. Koros, Engineering substructure morphology of asymmetric carbon molecular sieve hollow fiber membranes, *Carbon*, 76 (2014) 417–434.

[39] R. Kumar, C. Zhang, A.K. Itta, W.J. Koros, Highly permeable carbon molecular sieve membranes for efficient CO_2/N_2 separation at ambient and subambient temperatures, *Journal of Membrane Science*, 583 (2019) 9–15.

[40] O. Sanyal, S.T. Hicks, N. Bhuwania, S. Hays, M.G. Kamath, S. Karwa, R. Swaidan, W.J. Koros, Cause and effects of hyperskin features on carbon molecular sieve (CMS) membranes, *Journal of Membrane Science*, 551 (2018) 113–122.

[41] M.L. Jue, Y. Ma, R.P. Lively, Streamlined fabrication of asymmetric carbon molecular sieve hollow fiber membranes, *ACS Applied Polymer Materials*, 1 (2019) 1960–1964.

[42] A.B. Fuertes, D.M. Nevskaia, T.A. Centeno, Carbon composite membranes from Matrimid® and Kapton® polyimides for gas separation, *Microporous and Mesoporous Materials*, 33 (1999) 115–125.

[43] T.A. Centeno, A.B. Fuertes, Supported carbon molecular sieve membranes based on a phenolic resin, *Journal of Membrane Science*, 160 (1999) 201–211.

[44] W. Zhou, M. Yoshino, H. Kita, K.-i. Okamoto, Preparation and gas permeation properties of carbon molecular sieve membranes based on sulfonated phenolic resin, *Journal of Membrane Science*, 217 (2003) 55–67.

[45] M.A. Llosa Tanco, D.A. Pacheco Tanaka, A. Mendes, Composite-alumina-carbon molecular sieve membranes prepared from novolac resin and boehmite. Part II: Effect of the carbonization temperature on the gas permeation properties, *International Journal of Hydrogen Energy*, 40 (2015) 3485–3496.

[46] H. Wang, L. Zhang, G.R. Gavalas, Preparation of supported carbon membranes from furfuryl alcohol by vapor deposition polymerization, *Journal of Membrane Science*, 177 (2000) 25–31.

[47] T. Koga, H. Kita, K. Uemura, K. Tanaka, I. Kawafune, M. Funaoka, Structure and separation performance of carbon membranes from lignin-based materials, *Transactions of the Materials Research Society of Japan*, 34 (2009) 371–374.

[48] H. Li, Z. Song, X. Zhang, Y. Huang, S. Li, Y. Mao, H.J. Ploehn, Y. Bao, M. Yu, Ultrathin, molecular-sieving graphene oxide membranes for selective hydrogen separation, *Science*, 342 (2013) 95–98.

[49] L. Li, C. Song, H. Jiang, J. Qiu, T. Wang, Preparation and gas separation performance of supported carbon membranes with ordered mesoporous carbon interlayer, *Journal of Membrane Science*, 450 (2014) 469–477.

[50] C. Wang, L. Ling, Y. Huang, Y. Yao, Q. Song, Decoration of porous ceramic substrate with pencil for enhanced gas separation performance of carbon membrane, *Carbon*, 84 (2015) 151–159.

3

Carbon Membrane Characterization

Linfeng Lei

Department of Chemical Engineering, Norwegian University of Science and Technology

3.1 Introduction

Carbon molecular sieve (CMS) membranes are ultramicroporous inorganic membranes prepared by the carbonization of polymeric precursors, which consist primary of carbon atoms but usually also include a small amount of oxygen, nitrogen, and hydrogen. During carbonization, the polymeric precursors are transformed to carbon, at high temperatures and under vacuum or an inert gas environment (e.g., N_2 or Ar). The carbon membranes typically form a graphitic or turbostratic structure. Specifically, during the carbonization process, the entangled polymeric chains of precursors are gradually transformed to rigidly carbonized aromatic strands, which afterwards form organized plates [1, 2].

It is generally accepted that CMS membranes consist of a bimodal pore structure, with micropores (~7–20 Å) and ultramicropores (<7 Å). The micropores formed by the voids between aromatic carbon plates give the membranes high gas permeation flux while the ultramicropores produced by the slits between the highly aromatic strands control gas-pair selectivity [1, 3, 4]. Thus, it is necessary to investigate the fundamental properties of carbon membranes as these properties directly influence the separation performance. In this chapter, we describe different technologies for characterizing carbon membrane structure and separation performance. The evolution of membranes from a polymer precursor to a carbon matrix during the carbonization process, analyzed by thermogravimetric analysis (TGA), mass spectrometry (MS), and Fourier transform infrared spectroscopy (FTIR), is discussed in detail. The microstructure—morphology and pore size distribution—is usually characterized based on scanning electron microscopy (SEM), transmission electron microscopy (TEM), and physisorption. The type of hybridized carbon (sp^2 and sp^3) can be investigated by electron-based characterization techniques such as X-ray diffraction (XRD), Raman spectroscopy, X-ray photoelectron spectroscopy (XPS), and electron energy loss spectroscopy (EELS).

This chapter also describes the measurement of separation performance for carbon membranes used for gas separation.

3.2 Chemical Structural Evolution during Carbonization

During carbonization, the polymeric precursors are transformed to carbon at high temperature under vacuum or an inert gas environment (e.g., N_2, Ar, or CO_2). TGA, sometimes coupled with MS, is usually employed for investigating the evolution of the membranes during the carbonization process. He et al. [5] used a TGA–MS method to study the evolution of a deacetylated cellulose acetate precursor, as shown in Figure 3.1a. It was found that the degradation of the precursor started to occur at around 240 °C, and most of the weight loss occurred at approximately 300 °C to 650 °C, which was the main stage of polymer pyrolysis [5].

Volatile molecules such as H_2, CO_2, CO, H_2O, and CH_4 were detected and quantified by the coupled MS. According to the TGA–MS results, which showed that the generated CO and CO_2 were formed from the thermal cleavage of C–O and C–C bonds, the authors suggested that the carbonization mechanism might be dominated by the dehydration reaction. Following the ring opening, the volatiles were further eliminated from the glycoside links. Similarly, Ma et al. [1] investigated the amount of evolved CO_2 and H_2O during the carbonization of CMS membranes from a microporous polymer (PIM-1) under different H_2-containing atmospheres. It was found that the presence of H_2 in the pyrolysis environment can promote the formation of H_2O, by removing oxygen atoms from polymer chains, but inhibits formation

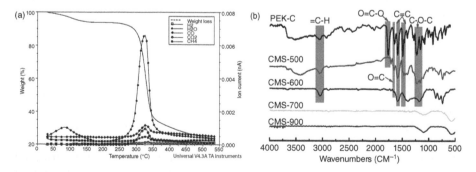

FIGURE 3.1
(a) TGA–MS analysis of the carbonization process from a deacetylated cellulose acetate precursor [5]; (b) FTIR spectra of PEK-C polymeric precursor and CMS membranes [6].

of CO_2. As a result, the H_2-assisted pyrolysis environment can lead to wider micropores than those formed in a pure Ar environment [1].

FTIR is a technique for identifying organic or inorganic functional groups from their characteristic infrared absorption peaks. Thus, it can be used to identify whether the polymer has been changed to a graphite-like structure after carbonization. Figure 3.1b shows the chemical evolution of carbon membranes prepared from a phenolphthalein-based cardo poly(arylene ether ketone) (PEK-C) precursor as a function of carbonization temperature. It indicates that CMS membranes made at low temperatures (500 °C and 600 °C) retain some degree of polymeric characteristics. However, the absorbance band signals were significantly weakened or disappeared when the carbonization temperature was raised. For example, the C–H stretching vibration peaks located at 3045 cm^{-1}, which correspond to the aromatic ring, disappeared when the carbonization temperature exceeded 700 °C. Thus, the final carbonization temperatures for preparation of CMS membranes should be higher than 700 °C [6].

3.3 Morphology of Carbon Membranes

Electron microscopy is a straightforward technique to determine the morphology of materials. Figure 3.2 illustrates the interaction between the incident electron beam and the sample in an electron microscope and the

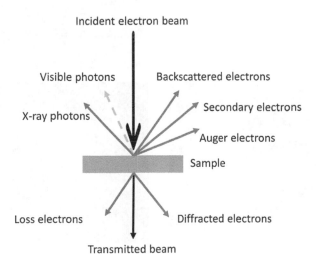

FIGURE 3.2
Illustration of the interaction between the incident electron beam and the sample in an electron microscope and the corresponding signals.

corresponding detected signals. The secondary and backscattered electrons are typically the basis of SEM images. In TEM, the transmitted electrons that pass through the sample are used. The main difference between SEM and TEM is that, in SEM, the degree of backscattering can identify the composition of the samples because heavy elements scatter electrons more efficiently. However, the contrast of TEM images, formed by the projection of transmitted electron beams that pass through the sample, rely solely on the density and thickness of the sample and are not influenced by its atomic structure.

SEM and TEM are commonly used to characterize membrane morphology. Figure 3.3 shows cross-sectional SEM images of different types of carbon membrane: self-supported film [7], supported carbon membrane [8], symmetric carbon hollow fiber [9], and asymmetric carbon hollow fiber [10]. It is important to determine the effective selective layer that works for sieving various gases, which can be easily detected using SEM. For example, the whole wall of the self-supported film and the symmetric hollow fiber are considered as the selective layer for the symmetric membrane, as shown in Figure 3.3a (75 μm) and Figure 3.3b (45 μm). For the supported membrane of an asymmetric hollow

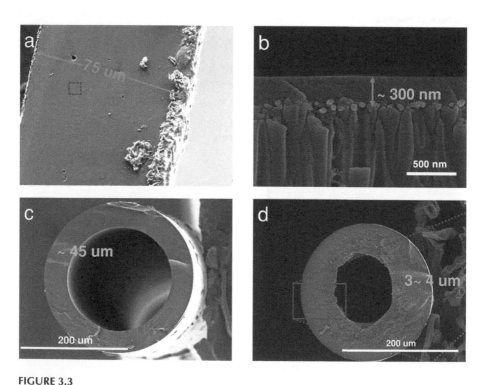

FIGURE 3.3
Cross-sectional SEM images of different types of carbon membranes: (a) self-supported film [7]; (b) supported carbon membrane [8]; (c) symmetric carbon hollow fiber [9]; and (d) asymmetric carbon hollow fiber [10].

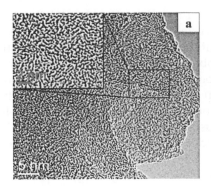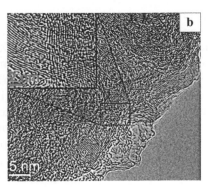

FIGURE 3.4
TEM images of carbon membranes carbonized at (a) 700 °C and (b) 900 °C [6].

fiber membrane, however, the selective layer is considered as the outer dense layer, as shown in Figure 3.3c (~300 nm) and Figure 3.3d (1~5 μm). If SEM is equipped with an energy-dispersive X-ray spectroscope (EDX or EDS), it can be used for elemental analysis. For instance, Tseng et al. [11] used an SEM–EDX line scanning analysis method to detect the penetration of the carbon into the TiO_2/Al_2O_3 composite support, with the aim of investigating the effects of interfacial adhesion between the CMS membrane and the composite support.

TEM has better imaging resolution (up to 0.2 nm) than SEM, so can be used to investigate the morphology and fine structure of carbon membranes [6, 7, 12]. Figure 3.4 presents a typical TEM image for a carbon membrane with a turbostratic carbon structure. Xu et al. found that carbon sheets tended to form ordered structures in the short range and to become dense as the carbonization temperature is increased from 700 °C to 900 °C [6]. For instance, a more ordered structure with parallel carbon sheets was observed in the carbon membrane carbonized at 900 °C (Figure 3.4b). Fourier transform of the high-resolution TEM provides a Debye–Scherrer ring that gives some information about the interplanar spacing of carbon membrane materials, as reported by Richter et al. [13].

3.4 Microstructure of Carbon Membranes

The CMS membranes typically possess a bimodal pore structure, with larger micropores (~7–20 Å) and smaller ultramicropores (<7 Å). In order to investigate the microstructure of the CMS membrane, different physical characterization methods such as XRD, XPS, EELS, Raman spectroscopy, and N_2 (CO_2) physisorption are employed.

XRD is one of the most frequently used techniques in material character-ization, being employed to identify crystalline phases of materials and to obtain the mean particle size. Thus, XRD patterns can provide information about the d-spacing of the CMS membrane, which represents the average inter-planar distance between the carbon sheets. Two broad peaks, around $2\theta = 23°$ and $2\theta = 45°$, normally exist in carbon membranes, which are char-acteristic of the amorphous carbonaceous structure (002) and graphitic-like (100) structure, respectively. Numerous studies have shown that the CMS membrane can be adjusted to a more ordered, denser crystalline structure by elevating carbonization temperature [1, 6, 14]. According to the XRD pattern, the d-spacing (*d*) for the samples can be obtained using Bragg's equation:

$$n\lambda = 2d \sin 2\theta \qquad\qquad (3.1)$$

where *n* is an integer, λ is the wavelength of the incident X-ray beam, and θ is the diffraction angle.

Figure 3.5a shows XRD patterns of the SBFDA-DMN polyimide precursor and its heat-derived CMS membranes [14]. It was reported that the d-spac-ing of the disordered graphitic planes (d_{002}) decreased slightly, from 4.0 Å to 3.8 Å, when the carbonization temperature was elevated from 800 to 1000 °C. As compared with the d_{002} of graphite (~3.4 Å), the shifted peaks implied that the CMS membrane was becoming a more ordered graphitic structure with smaller pores. Moreover, the calculated *d*-spacing of the (100) plane, d_{100}, around 2.0–2.1 Å, located at $2\theta = 44°$, indicates the average space between the neighboring carbon atoms within graphene planes [14].

The type of defects in the graphitic layers of the CMS membranes, such as sp²- and sp³-hybridized carbon, can be detected by Raman spectroscopy,

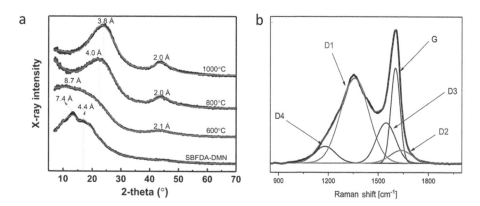

FIGURE 3.5
(a) XRD patterns of the SBFDA-DMN polyimide precursor and its heat-derived CMS mem-branes [14]; (b) Raman spectrum of the CMS membrane made from polyimide precursor [15].

which is most sensitive to highly symmetric covalent bonds. As shown in Figure 3.5b, a Raman spectrum of a CMS membrane gives a typical amorphous carbon, which has a G band (located at about 1600 cm^{-1}) that is the characteristic peak of the ideal graphitic vibration mode (E_{2g}-symmetry) and a D band (located at about 1380 cm^{-1}) that is the disordered graphite peak (graphene layer edges, A_{1g}-symmetry). Moreover, the spectrum can be deconvoluted into five bands: D1, D2, D3, D4, and G. The D2 band corresponds to the graphitic lattice vibrations mode with E_{2g}-symmetry but involving isolated graphene layers [4]. The D3 and D4 bands are assigned to highly defective carbonaceous materials [15]. The type of hybridized carbon is then distinguished by calculating the ratio of D1 to D2 intensity. It was suggested that the sp^3 hybridization carbon defect occurs when the ratio is about 13, whereas vacancy-like defects dominate when the ratio is close to 7 [16]. For example, the ratio of D1 to D2 is calculated as 13.1 in Figure 3.5b, indicating that the prepared CMS membrane contains mostly sp^3-hybridized carbon.

XPS, which is a surface characterization analysis method that measures a depth of up to 5 nm from the sample surface, gives information about the elemental composition and the chemical state of the elements. It provides another method to identify the carbon structure. The C 1s XPS spectrum for a CMS membrane fabricated from a PIM-1 precursor was reported by Ma et al. [1]. The spectrum can be deconvoluted into three peaks. The peaks located at ~284.4 eV and 285.5 eV mainly correspond to the C=C bond (sp^2 carbon) and C–C bond (sp^3), respectively, and the peak at ~289 eV is assigned to C–O or C–N. Ma et al. [1] also investigated the C 1s XPS spectra of CMS membranes that were fabricated under different carbonization environments. The authors suggested that sp^2-hybridized carbon can be converted to sp^3-hybridized carbon when H_2 is introduced during the carbonization or a higher carbonization temperature was applied. For example, the CMS membrane carbonized under 4 % H_2/Ar presented an sp^3/sp^2 ratio of 0.65, higher than that of the CMS membrane that was carbonized under pure Ar, 0.24.

Richter et al. reported that EELS can be used to distinguish sp^2 and sp^3 carbons [13]. For carbon materials, the EELS spectrum K-edge represents electrons transitioning from 1s states to antibonding π^* states (π^* band) or antibonding σ^* states (σ^* band) [17]. Specifically, sp^3 carbon orbitals do not possess π states so exhibit only one major feature edge at about 295 eV (σ^* band), whereas the EELS spectrum of sp^2 carbon exhibits two major features, at about 285 eV (π^* band) and 295 eV (σ^* band). Richter et al. [13] compared different carbon materials: carbon membrane, graphite, diamond, and amorphous carbon. It was found that the carbon membrane exhibited a stronger π^* band than amorphous carbon, which indicated that it contained a certain amount of sp^2-hybridized carbon. The authors then confirmed using a combination of EELS, Raman, and XPS that the carbon membrane was a mix of about two-thirds sp^2-hybridized carbon and one-third sp^3-hybridized carbon [13].

N_2 and CO_2 are often used as probe molecules to estimate the pore size distribution of CMS membranes by adsorption measurements. N_2 adsorption at 77 K is widely used for the investigation of porous materials. Figure 3.6a gives an example of the pore size distribution of a CMS membrane, obtained from N_2 adsorption and reported by Hazazi et al. [14]. Generally, a higher carbonization temperature will narrow both the micropores and ultramicropores of CMS membranes. As shown in Figure 3.6a, the pore size distribution shifts to smaller pores when carbonized at higher temperature. Moreover, under an extremely high temperature, 1000 °C, it was not detectable by N_2 physisorption experiments. To overcome the problem that N_2 (3.64 Å) is difficult to diffuse into the ultramicropores in CMS membranes (<7 Å), the smaller CO_2 molecule (3.3 Å) is normally employed at 273 K for the assessment of the ultramicropores. As shown in Figure 3.6b, the CMS membranes shows a pore size distribution of ultramicropores around 4–7 Å.

The structural properties, such as micropore volume and average micropore width, can also be evaluated by high-pressure CO_2 adsorption measurement at 298 K [9, 18, 19], which is based on the Dubinin–Radushkevich (DR) equation [20] (Equation 3.2) and the Stoeckli equation [21] (Equation 3.3).

$$\frac{w}{w_0} = \exp\left(-\left(\frac{RT \ln\left(\frac{p_0}{p}\right)}{\beta E_0}\right)^2\right) \tag{3.2}$$

$$L_0 = \frac{10.8\,(\text{nm}\cdot\text{kJ}/\text{mol})}{E_0 - 11.4\,(\text{kJ}/\text{mol})} \tag{3.3}$$

where w and w_0 are the gas volume adsorbed at pressure p and the micropore volume of the CMS membranes, respectively; p_0 is the saturation pressure of CO_2 at 298 K; β is the affinity coefficient, a characteristic of the adsorption capacity; E_0 and L_0 correspond to the adsorption activation energy and average micropore size, respectively, of a CMS sample. Equation 3.2 can be rewritten as,

$$\text{Ln}\,w = \text{Ln}\,w_0 - \left(\frac{RT}{\beta E_0}\right)^2 \ln\left(\frac{p_0}{p}\right)^2 \tag{3.4}$$

Thus, there is a linear relationship between the adsorbed gas volume and the testing pressure. The micropore volume and adsorption activation energy are deduced by the intercept and slope, respectively. The DR equation

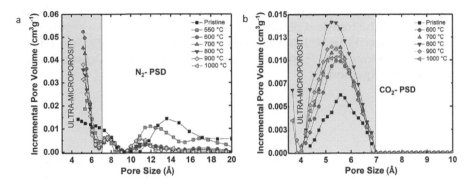

FIGURE 3.6
Pore size distribution for CMS membrane, obtained from (a) N_2 physisorption and (b) CO_2 physisorption [14].

is modified to the Dubinin–Astakhov equation (Equation 3.5) [22] when the DR equation does not describe a linear curve.

$$\frac{w}{w_0} = \exp\left(-\left(\frac{RT \ln\left(\frac{p_0}{p}\right)}{\beta E_0}\right)^n\right) \qquad (3.5)$$

where n is an adjustable parameter that can be obtained by modeling regression.

The true density (ρ_b) of CMS materials is determined by buoyancy measurement with non-absorbable helium gas [19]. The bulk density (ρ_s) of samples is calculated using equation 3.6.

$$\frac{1}{\rho_b} = \frac{1}{\rho_s} + w_o \qquad (3.6)$$

3.5 Gas Separation Measurement

Carbon membrane separation performance can be determined by gas permeation measurements with single or mixed gas. The single gas permeation measurement is widely conducted on a constant permeate volume method with a variable-pressure system. It may take several minutes to several hours to ensure that the transient phase of diffusion is complete and a steady state has been achieved (dp/dt tends to a constant). The gas permeability, P (barrer, where 1 barrer = 10^{-10} cm³ (STP)·cm/(cm²·s·cmHg)) is calculated using the following equation:

$$P = \frac{273 \times 10^7 V \cdot y_i}{76T \cdot A \cdot \Delta p} \bullet \left(\frac{dp}{dt}\right) \tag{3.7}$$

where V is the collection volume (cm³) that can be measured with a pre-calibrated permeation cell reported elsewhere [23, 24], dp/dt is the collection volume pressure increase rate (mbar/s), l and A are thickness (cm) and total effective area (cm²) of the membrane sample, respectively, Δp (bar) is the pressure drop across the membrane, and T the experimental temperature (K). In this work, the ideal selectivity is defined as the ratio of the pure gas permeability values, which is evaluated as follows:

$$\alpha_{i/j} = \frac{P_i}{P_j} \tag{3.8}$$

For mixed gas separation measurements, a permeation cell and a gas chromatograph (GC) were combined to analyze the gas composition and calculate the permeability of the gas component. The separation factor can be calculated using the following equation:

$$\alpha_{i/j} = \frac{y_i / y_j}{x_i / x_j} \tag{3.9}$$

where x_i is the feed composition of component i, and y_i the molar fraction of component i in the permeate stream, respectively, which can be measured by GC. Figure 3.7 gives an example of the schematic diagram for a high-pressure CO_2/CH_4 mixed gas test. The flow and pressure of feed gas (CO_2/CH_4) are controlled by mass flow controller (MFC 1) and back pressure controller (BPC 1), respectively. N_2 is used as sweep gas flow and is controlled by MFC 2 and BPC 2, and the permeate flow including sweep gas is measured by the mass flow meter (MFM 1). The gas composition at the permeate side is analyzed by a GC. The gas temperature in the feed, retentate, and permeate sides are measured by temperature indicators (TI1-3).

To obtain the diffusivity (D) of the carbon membranes, a time-lag method is widely used:

$$D = \frac{l^2}{6\theta} \tag{3.10}$$

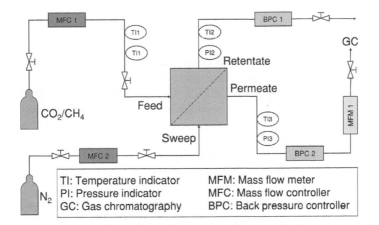

FIGURE 3.7
Schematic diagram for mixed gas permeation testing [9].

in which l is thickness of the selective layer of the tested carbon membrane and θ the permeation time lag as shown in Figure 3.8. Subsequently, the sorption coefficient (S) can be calculated by the equation $P = D \times S$. The sorption coefficient can also be determined by sorption measurement.

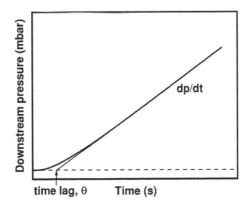

FIGURE 3.8
Schematic showing determination of permeation time lag (θ).

3.6 Conclusions

In this chapter the most common characterization techniques used for carbon membranes are discussed, including TGA, MS, FTIR, SEM, TEM, XRD, XPS, Raman spectroscopy, and EELS. Using these characterization techniques, we can investigate a carbon membrane from a fundamental perspective that aims to understand the microstructure at the molecular level where the separation happens. Although some of the characterizations, such XPS, SEM, and TEM, require an ultrahigh vacuum situation, which may exhibit different behavior from realistic conditions, the obtained results can help us to improve the separation performance of the membranes.

It is important to combine all characterization techniques that reveal credible and usable information on materials. Two key factors are commonly used to describe the separation performances of membranes, permeability and selectivity (or separation factor), which indicate the productivity and separation efficiency of the separation process, respectively. They are determined by single gas and mixed gas permeation tests. This chapter has also described how to conduct a permeation test.

Acknowledgments

The authors would like to thank the Research Council of Norway for funding this work through the PETROMAKS 2 program in the CO2Hing project (#267615).

List of Acronyms

BPC	Back pressure controller
CMS	Carbon molecular sieve
DR	Dubinin–Radushkevich
EDX	Energy-dispersive X-ray spectroscopy
EELS	Electron energy loss spectroscopy
FTIR	Fourier transform infrared spectroscopy
GC	Gas chromatograph
MFC	Mass flow controller
MFM	Mass flow meter
MS	Mass spectrometry

PEK-C Poly (arylene ether ketone)
SEM   Scanning electron microscopy
STP   Standard temperature and pressure
TGA   Thermogravimetric analysis
TEM   Transmission electron microscopy
XPS   X-ray photoelectron spectroscopy
XRD   X-ray diffraction

Nomenclature

$\alpha_{i/j}$ gas selectivity of i to j
β affinity coefficient
θ diffraction angle (°) in Equation 3.1, and permeation time lag in Equation 3.10 (s)
λ wavelength of X-ray (nm)
ρ_b true density (g/cm^3)
ρ_s bulk density (g/cm^3)
A effective membrane area (cm^2)
D diffusivity (cm^2/s)
E_0 adsorption activation energy
l effective membrane thickness (μm)
L average micropore size (nm)
p pressure (bar)
P gas permeability (barrer) (1 Barrer = 0.33 x 10^{-15} mol m m^{-2} s^{-1} Pa^{-1})
p_0 saturation pressure (bar)
S sorption coefficient (cm^3 [STP]/(cm^2·s cmHg))
T temperature (K)
V volume (cm^3)
w gas volume adsorbed (cm^3/g)
w_0 micropore volume (cm^3)
x_i mole fraction of i in feed
y_i mole fraction of i in permeate

References

[1] Y. Ma, M.L. Jue, F. Zhang, R. Mathias, H.Y. Jang, R.P. Lively, Creation of well-defined "mid-sized" micropores in carbon molecular sieve membranes, *Angewandte Chemie*, 131 (2019) 13393–13399.

[2] M. Rungta, G.B. Wenz, C. Zhang, L. Xu, W. Qiu, J.S. Adams, W.J. Koros, Carbon molecular sieve structure development and membrane performance relationships, *Carbon*, 115 (2017) 237–248.

[3] C. Zhang, W.J. Koros, Ultraselective carbon molecular sieve membranes with tailored synergistic sorption selective properties, *Advanced Materials*, 29 (2017) 1701631.

[4] W. Qiu, J. Vaughn, G. Liu, L. Xu, M. Brayden, M. Martinez, T. Fitzgibbons, G. Wenz, W.J. Koros, Hyperaging tuning of a carbon molecular-sieve hollow fiber membrane with extraordinary gas-separation performance and stability, *Angewandte Chemie International Edition*, 58 (2019) 11700–11703.

[5] X. He, J.A. Lie, E. Sheridan, M.-B. Hagg, Preparation and characterization of hollow fiber carbon membranes from cellulose acetate precursors, *Industrial & Engineering Chemistry Research*, 50 (2011) 2080–2087.

[6] R. Xu, L. He, L. Li, M. Hou, Y. Wang, B. Zhang, C. Liang, T. Wang, Ultraselective carbon molecular sieve membrane for hydrogen purification, *Journal of Energy Chemistry*, 50 (2020) 16–24.

[7] Z. Yang, W. Guo, S.M. Mahurin, S. Wang, H. Chen, L. Cheng, K. Jie, H.M. Meyer, D.-e. Jiang, G. Liu, W. Jin, I. Popovs, S. Dai, Surpassing Robeson upper limit for CO_2/N_2 Separation with fluorinated carbon molecular sieve membranes, *Chem*, 6 (2020) 631–645.

[8] J. Hou, H. Zhang, Y. Hu, X. Li, X. Chen, S. Kim, Y. Wang, G.P. Simon, H. Wang, Carbon nanotube networks as nanoscaffolds for fabricating ultrathin carbon molecular sieve membranes, *ACS Applied Materials & Interfaces*, 10 (2018) 20182–20188.

[9] L. Lei, A. Lindbråthen, M. Hillestad, M. Sandru, E.P. Favvas, X. He, Screening cellulose spinning parameters for fabrication of novel carbon hollow fiber membranes for gas separation, *Industrial & Engineering Chemistry Research*, 58 (2019) 13330–13339.

[10] N. Bhuwania, Y. Labreche, C.S.K. Achoundong, J. Baltazar, S.K. Burgess, S. Karwa, L. Xu, C.L. Henderson, P.J. Williams, W.J. Koros, Engineering substructure morphology of asymmetric carbon molecular sieve hollow fiber membranes, *Carbon*, 76 (2014) 417–434.

[11] H.-H. Tseng, C.-T. Wang, G.-L. Zhuang, P. Uchytil, J. Reznickova, K. Setnickova, Enhanced H_2/CH_4 and H_2/CO_2 separation by carbon molecular sieve membrane coated on titania modified alumina support: Effects of TiO_2 intermediate layer preparation variables on interfacial adhesion, *Journal of Membrane Science*, 510 (2016) 391–404.

[12] S.C. Rodrigues, M. Andrade, J. Moffat, F.D. Magalhães, A. Mendes, Carbon membranes with extremely high separation factors and stability, *Energy Technology*, 7 (2019) 1801089.

[13] H. Richter, H. Voss, N. Kaltenborn, S. Kämnitz, A. Wollbrink, A. Feldhoff, J. Caro, S. Roitsch, I. Voigt, High-flux carbon molecular sieve membranes for gas separation, *Angewandte Chemie International Edition*, 56 (2017) 7760–7763.

[14] K. Hazazi, X. Ma, Y. Wang, W. Ogieglo, A. Alhazmi, Y. Han, I. Pinnau, Ultraselective carbon molecular sieve membranes for natural gas separations based on a carbon-rich intrinsically microporous polyimide precursor, *Journal of Membrane Science*, 585 (2019) 1–9.

[15] P.H.T. Ngamou, M.E. Ivanova, O. Guillon, W.A. Meulenberg, High-performance carbon molecular sieve membranes for hydrogen purification and pervaporation dehydration of organic solvents, *Journal of Materials Chemistry A*, 7 (2019) 7082–7091.

[16] A. Eckmann, A. Felten, A. Mishchenko, L. Britnell, R. Krupke, K.S. Novoselov, C. Casiraghi, Probing the nature of defects in graphene by Raman spectroscopy, *Nano Letters*, 12 (2012) 3925–3930.

[17] J. Gao, Y. Wang, H. Wu, X. Liu, L. Wang, Q. Yu, A. Li, H. Wang, C. Song, Z. Gao, M. Peng, M. Zhang, N. Ma, J. Wang, W. Zhou, G. Wang, Z. Yin, D. Ma, Construction of a sp^3/sp^2 Carbon interface in 3D N-doped nanocarbons for the oxygen reduction reaction, *Angewandte Chemie International Edition*, 58 (2019) 15089–15097.

[18] S.C. Rodrigues, M. Andrade, J. Moffat, F.D. Magalhães, A. Mendes, Preparation of carbon molecular sieve membranes from an optimized ionic liquid-regenerated cellulose precursor, *Journal of Membrane Science*, 572 (2019) 390–400.

[19] X. He, M.-B. Hägg, Structural, kinetic and performance characterization of hollow fiber carbon membranes, *Journal of Membrane Science*, 390–391 (2012) 23–31.

[20] M.M. Dubinin, Generalization of the theory of volume filling of micropores to nonhomogeneous microporous structures, *Carbon*, 23 (1985) 373–380.

[21] F. Stoeckli, A. Slasli, D. Hugi-Cleary, A. Guillot, The characterization of microporosity in carbons with molecular sieve effects, *Microporous and Mesoporous Materials*, 51 (2002) 197–202.

[22] A. Gil, P. Grange, Application of the Dubinin-Radushkevich and Dubinin-Astakhov equations in the characterization of microporous solids, *Colloids and Surfaces A: Physicochemical and Engineering Aspects*, 113 (1996) 39–50.

[23] W.-H. Lin, R.H. Vora, T.-S. Chung, Gas transport properties of 6FDA-durene/1, 4-phenylenediamine (pPDA) copolyimides, *Journal of Polymer Science Part B: Polymer Physics*, 38 (2000) 2703–2713.

[24] J.A. Lie, Synthesis, performance and regeneration of carbon membranes for biogas upgrading-a future energy carrier, Department of Chemical Engineering, Norwegian University of Science and technology, Trondheim, 2005.

4

Carbon Membrane Transport Mechanisms

Xuezhong He

Department of Chemical Engineering, Norwegian University of Science and Technology

Department of Chemical Engineering, Guangdong Technion Israel Institute of Technology (GTIIT)

4.1 Introduction

Gas separation by membrane technology has been subject to rapidly growing interest as an alternative to traditional gas separation methods within the past 30 years. The membrane technique is a low-cost, energy-efficient process with high flexibility and no requirement for chemicals and solvents. However, the greatest challenges for membrane technology are the requirement for pretreatment of the feed gas stream, and membrane lifetime. Membranes for gas separation are characterized to separate a particular gas mixture.

The membrane is a selective barrier that has the ability to transport one component more readily than others as a result of differences in physical and/or chemical properties. Gas transport through a membrane is a result of the driving force of the trans-membrane pressure. However, there are different transport mechanisms for gas molecules passing through carbon membranes. This chapter describes the three main transport mechanisms: molecular sieving, selective surface flow and Knudsen diffusion.

General mass transport through the membrane and process parameters influencing the process are also discussed, and will be used to guide membrane modeling and process simulation for specific applications of carbon membranes.

4.2 General Gas Transport Model

The most common configuration for gas transport through a membrane is the perfect mixing model, which is shown in Figure 4.1. The transport of gas molecules through membranes can be described by Fick's first law, which gives, for the unidimensional flux J_i for component i through a membrane,

$$J_i = -D_i \frac{dc_i}{dx_i} \tag{4.1}$$

where D_i is the diffusion coefficient for component i and $\frac{dc_i}{dx_i}$ is the driving force. For ideal systems, where gas solubility is independent of concentration and can be described by Henry's law ($c = S \times p$), the flux of component i, J_i (m^3 (STP)/($m^2 \cdot h$)), can be modified by,

$$J_i = \frac{P_i}{l} \Delta p_i = \frac{P_i}{l} \left(p_H x_{i,F} - p_L y_{i,P} \right) \tag{4.2}$$

where P_i is the permeability (barrer) for component i [1], l is membrane thickness (m), Δp is driving force, p is total gas pressure (subscripts H and L

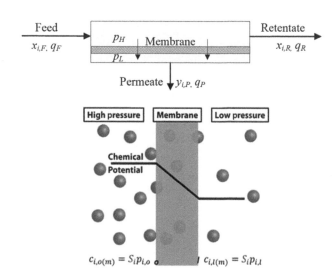

FIGURE 4.1
Schematic diagram for a gas membrane separation process.

represent the high- and low-pressure side, respectively) (bar) and $x_{i,F}$ and $y_{i,P}$ are the mole fractions of component i on the feed and permeate sides, respectively. The ideal selectivity is defined as,

$$\alpha = \frac{P_i}{P_j} \tag{4.3}$$

However, for a gas mixture separation process, process selectivity (or the separation factor) α is described as follows,

$$\alpha_{i/j} = \frac{y_{i,P} / y_{j,P}}{x_{i,F} / x_{j,F}} \tag{4.4}$$

Stage-cut, which is an important economic index, is defined as $\theta = \frac{q_P}{q_F} \times 100\%$ (q is gas flow rate). The stage-cut is typically higher in a real separation process compared with laboratory-scale gas permeation tests (which usually operate at a very low stage-cut of $< 1\%$). Therefore, choosing a correct stage-cut is essential as it directly influences the purity and yield of products.

4.3 Transport through Carbon Membranes

The ability of ultramicroporous carbon membranes to separate a gas mixture depends on membrane pore size, the physiochemical properties of the gases and the surface properties of the membrane pores. The pore size of a carbon fiber for gas separation is usually within the range of 3.5 to 10 Å, depending on the membrane preparation conditions during carbonization or post treatment (post oxidation or chemical vapor deposition). There are basically three transport mechanisms for carbon membranes, as listed below [2]; see also the illustration in Figure 4.2:

1. Knudsen diffusion; the square root of the ratio of gas molecular weights will give the separation factor.
2. Selective surface diffusion, governed by selective adsorption of the larger non-ideal components on the pore surface, hence the smaller components are retained.
3. Molecular sieving; smaller molecules will permeate through carbon membranes, with larger molecules being retained or permeating at a much lower rate.

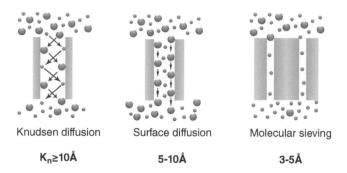

Knudsen diffusion   Surface diffusion   Molecular sieving

$K_n \geq 10 \text{Å}$ 5-10Å   3-5Å

FIGURE 4.2
Illustration of different transport mechanisms through carbon membranes.

4.3.1 Knudsen Diffusion

For Knudsen diffusion, the lower limit for pore diameter has usually been set to $d_p > 20$ Å [3]. However, Gilron and Soffer [4] have discussed thoroughly how Knudsen diffusion may contribute to transport in even smaller pores and, from a model considering pore structure, shown that contributions to transport through one specific fiber may come from both activated transport and Knudsen diffusion. It may therefore be difficult to know exactly when transport due to Knudsen diffusion is taking place. One way to approach this problem is to calculate the Knudsen number, N_{Kn}, for a system, which is λ/d_p, where λ is the mean free path. If $N_{Kn} \geq 10$, then the separation can be assumed to take place according to Knudsen diffusion [5]. If the preparation of carbon membranes has been unsuccessful, one may get Knudsen diffusion. Gas flux based on Knudsen diffusion is similar to Fick's law, and can be described as follows,

$$J_i = -\frac{D_k}{RT}\frac{dp_i}{dx} \tag{4.5}$$

$$D_k = \left(\frac{2}{3}\right) r_p \bar{v} \tag{4.6}$$

$$\bar{v} = \sqrt{\frac{8RT}{\pi M}} \tag{4.7}$$

where r_p is pore radius, v is the average velocity of gas molecules, M is gas molecular weight and T is the operating temperature. Gas selectivity is then dependent on the ratio of molecular weight of gas molecules:

$$\alpha_{ij} = \sqrt{\frac{M_j}{M_i}} \tag{4.8}$$

4.3.2 Selective Surface Flow

The driving force for separation according to selective surface flow is basically the difference in the concentration of diffusing components in the adsorbed phase. This means that a high driving force can be attained even with a small partial pressure difference for permeating components. The larger molecules (more condensable, e.g., hydrocarbon) in a gas mixture will be selectively adsorbed, so the smaller molecules will be retained due to reduced pore size. The pore size at which selective surface flow is expected to take place is about 5 Å $< d_p <$ 10 Å; or up to 3 × (diameter of molecule) [3]. The transport of gas molecules through a carbon membrane can also be described by Fick's first law as given in Equation 4.1, and the activated diffusion is described by an Arrhenius-type equation:

$$D_a = D_0 \cdot \exp\left(-E_d / RT\right) \tag{4.9}$$

where E_d is the activation energy for diffusion. Now, if Henry's law is assumed to apply, the integrated flux equation may be written as in Equation 4.10:

$$J_a = \frac{\Delta p}{RT \cdot l} D_0 \cdot \exp\left\{\frac{-\left(E_{a,S} - E_{ads}\right)}{RT}\right\} = \frac{\Delta p}{RT \cdot l} D_0 \cdot \exp\left(\frac{-\Delta E_S}{RT}\right) \tag{4.10}$$

ΔE_S, the difference between transport activation energy and adsorption energy, may be positive or negative. When $\Delta E_S < 0$, transport due to selective surface flow will be enhanced with increasing temperature; with $\Delta E_S > 0$ it will decrease.

4.3.3 Molecular Sieving

Molecular sieving is the dominating transport mechanism in carbon membranes; this has also given them the name carbon molecular sieve (CMS) membranes. The pore size is usually within the range of a few (3–5) Å (see Figure 4.2). The dimensions of a molecule are usually described either with the Lennard-Jones radius or the van der Waals radius. The sorption selectivity has little influence on separation when molecular sieving is considered. Equation 4.9 is still valid for activated transport, but now attention should be drawn to the pre-exponential term, D_0 ($D_0 = \frac{e\lambda^2 kT}{h} \exp\left(\frac{S_{a,d}}{R}\right)$) [6]. Thus, the flux for single component can be expressed as,

$$J_a = \frac{\Delta P}{RTL} D_0 \cdot \exp\left(\frac{-E_{a,MS}}{RT}\right) \tag{4.11}$$

Here, $E_{a,\,MS}$ is the activation energy for diffusion in a molecular sieving process for CMS membranes. Nguyen et al. reported that CMS membranes present a reasonable sieving effect for gas molecules with different kinetic diameters, which suggests that CMS membranes are predominantly microporous with no major contribution from Knudsen diffusion or viscous flow in their overall mass transfer [7].

4.4 Process Parameter Influences

The gas separation performance for a given membrane system mainly depends on membrane material properties (i.e., gas permeability and selectivity). However, the operating parameters (i.e., pressure, temperature) for a specific process can also affect the membrane separation performance.

4.4.1 Effect of Pressure

One of the most important parameters is the pressure ratio across a membrane, which is defined as the ratio between feed and permeate pressures. Component i can only transport through membrane when the partial pressure on the feed side (p_H) is higher than that on the permeate side (p_L), as indicated in Equation 4.12.

$$x_{i,F}p_H > y_{i,P}p_L \quad \text{or} \quad \frac{y_{i,P}}{x_{i,F}} < \frac{p_H}{p_L} = \phi \tag{4.12}$$

It is also found from Equation 4.12 that enrichment of component i can never exceed the pressure ratio regardless of membrane selectivity [8]. The relationship between pressure ratio and membrane selectivity can be derived from Equations 4.2 and 4.4 [9,10].

$$y_{i,p} = \frac{\phi}{2}\left[x_{i,F} + \frac{1}{\phi} + \frac{1}{\alpha - 1} + \sqrt{\left(x_{i,F} + \frac{1}{\phi} + \frac{1}{\alpha - 1} \right)^2 - 4\frac{\alpha x_{i,F}}{(a-1)\phi}} \right] \tag{4.13}$$

If membrane selectivity (α) is much larger than pressure ratio (Φ) (i.e., $\alpha \gg \Phi$), Equation 4.13 can be simplified as,

$$y_{i,p} = x_{i,F}\phi \tag{4.14}$$

This is normally called the pressure-ratio-limited region, and membrane separation performance is mainly determined by the pressure ratio across membranes while selectivity has only a minor effect. However, if membrane selectivity is much smaller than pressure ratio ($\alpha \ll \phi$), Equation 4.13 becomes,

$$y_{i,P} = \frac{\alpha x_{i,F}}{1 - x_{i,F}(1-\alpha)} \tag{4.15}$$

This is the membrane-selectivity-limited region; membrane separation performance is mainly controlled by membrane selectivity while the pressure ratio has a minor effect. In between these two extremes, both pressure ratio and membrane selectivity will influence membrane system performance.

An example of the dependence of the permeate concentration on pressure ratio and selectivity was reported by Paul et al. [9]. The pressure ratio is very important for gas separation processes at the industrial scale due to practical limitations. Achieving a high pressure ratio by compressing feed gas to high pressure or applying a high vacuum to the permeate side will significantly increase energy costs. Therefore, the practical pressure ratios are typically in the range of 5–20 [8].

4.4.2 Effect of Temperature

Gas transport through carbon membranes may be considered as an activated process, which can typically be described by an Arrhenius equation [11],

$$P = P_0 \exp\left(\frac{-E_p}{RT}\right) = S_0 D_0 \exp\left(-\frac{\Delta H_s + E_d}{RT}\right) \tag{4.16}$$

where P_0, S_0 and D_0 are the preexponential factors and ΔH_s and E_d are the heat of solution and activation energy for diffusion, respectively. Temperature has a significant effect on gas permeability. For small, non-interactive gas molecules, the effect of temperature on gas permeability is mainly determined by diffusivity, as temperature has a minor influence on the solubility coefficient. However, for large gas molecules, the temperature effects on solubility coefficient and diffusivity are in opposition and thus gas permeability will be determined by the dominant parameter. Lei et al. reported the temperature dependence of the separation performance of a cellulose-based carbon membrane operating at a feed pressure of 8 bar [12]. When the temperature was increased from 25 to 60 °C, both CO_2 and CH_4 permeabilities increased, whereas the separation factor decreased, as shown in Figure 4.3. They also reported that increasing the operating temperature enhances the CO_2 diffusion coefficient but CO_2 adsorption in the carbon matrix decreases. Overall, it causes a relatively slower increase of CO_2 permeability compared with that of CH_4. The apparent transport activation energies, calculated from

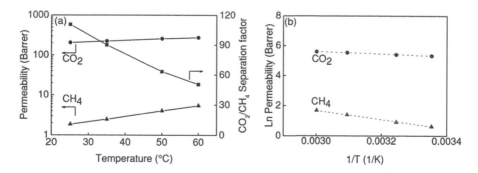

FIGURE 4.3
Effects of operation temperature on separation performance of a cellulose-based carbon hollow fiber membrane. (a) Temperature dependence of CO_2 permeability and CO_2/CH_4 separation factor; (b) Arrhenius plots according to CO_2 and CH_4 permeability [12].

the Arrhenius equation ($P = P_0 e^{-\frac{E_a}{RT}}$), are 6.7 and 25.4 kJ/mol for CO_2 and CH_4, respectively. The larger E_a of CH_4 indicates a lower permeability at the same operating temperature but a more significant effect of temperature.

4.5 Summary

Depending on the pore size of the carbon membrane, different transport mechanisms may dominate membrane separation processes. Carbon membranes with a selective surface flow mechanism may be applied to separation of heavy hydrocarbons, whereas carbon molecular sieve membranes are particularly useful for the separation of small gas molecules that are alike in size such as H_2–CO_2, CO_2–CH_4, olefin–paraffin, organic solvents etc. (see details in Part 2: Carbon Membrane Applications).

Nomenclature

Symbol	Explanation	Unit/value
Latin characters		
c	concentration	mol m^3
D	diffusion coefficient	m^2 S^{-1}
E	activation energy	kJ mol^{-1}

Symbol	Explanation	Unit/value
h	Planck constant	6.626×10^{-34} J s
J	flux	mol m^{-2} s^{-1}
k	Boltzmann constant	1.381×10^{-23} J K^{-1}
l	Membrane thickness	m
M	Molecular weight	g mol^{-1}
N	Knudsen number	–
P	permeability	Barrer (1 Barrer = 0.33 $\times$ 10^{-15} mol m m^{-2} s^{-1} Pa^{-1})
p	pressure	Bar, or mbar
q	flow rate	m^3 h^{-1}
R	molar gas constant	8.314 J K^{-1}mol^{-1}
r_p	pore radius	m
S	entropy; solubility	J K^{-1}; mol m^{-3} bar^1
T	temperature	K
v	gas velocity	m s^{-1}
x	mole composition	–
y	mole composition	–

Greek characters

Δ	delta (finite difference)	–
Π	permeation number	–
Φ	pressure ratio	–
α	selectivity	–
θ	stage-cut	–
λ	mean free path	m

Subscripts

A	activation
d	diffusion
F	feed side
H, L	high, low pressure side
i, j	component i, j
K	Knudson
P	permeate side
p	preexponential
s	sorption
0	usually presents pre-exponential

References

[1] W.J. Koros, H. Ma, T. Shimidzu, Terminology for membranes and membrane processes (IUPAC Recommendation 1996), *J Membr Sci*, 120 (1996) 149–159.

[2] M.-B. Hagg, J.A. Lie, A. Lindbrathen, carbon molecular sieve membranes. A promising alternative for selected industrial applications, . A promising alternative for selected industrial applications, *Ann NY Acad Sci*, 984 (2003) 329–345.

[3] M.B. Rao, S. Sircar, Performance and pore characterization of nanoporous carbon membranes for gas separation, *J Membr Sci*, 110 (1996) 109–118.

[4] J. Gilron, A. Soffer, Knudsen diffusion in microporous carbon membranes with molecular sieving character, *J Membr Sci*, 209 (2002) 339–352.

[5] C.J. Geankoplis, *Transport Processes and Unit Operations*, 3rd ed., Prentice-Hall, Englewood Cliffs, NJ, 1993.

[6] S. Glasstone, K.J. Laidler, H. Eyring, *The Theory of Rate Processes*, 1st ed., McGraw-Hill Book Co., New York, 1941.

[7] C. Nguyen, D.D. Do, K. Haraya, K. Wang, The structural characterization of carbon molecular sieve membrane (CMSM) via gas adsorption, *J Membr Sci*, 220 (2003) 177–182.

[8] R.W. Baker, *Membrane Technology and Applications*, 2nd ed., Wiley, Chichester, 2004.

[9] D.R. Paul, J.P. Jampol'skij, *Polymeric Gas Separation Membranes*, CRC Press, Boca Raton, FL, 1994.

[10] R. Baker, *Membrane Technology and Applications*, 2nd ed., McGraw-Hill, 2004.

[11] M. Mulder, *Basic Principles of Membrane Technology*, Kluwer Academic Publishers, Dordrecht, the Netherlands, pp. 496–498.

[12] L. Lei, A. Lindbråthen, M. Hillestad, M. Sandru, E.P. Favvas, X. He, Screening cellulose spinning parameters for fabrication of novel carbon hollow fiber membranes for gas separation, *Ind Eng Chem Res*, 58 (2019) 13330–13339.

Part 2

Carbon Membrane Applications

5

Carbon Membranes for Biogas Upgrading

Arne Lindbråthen

Department of Chemical Engineering, Norwegian University of Technology and Science (NTNU)

5.1 Background

Biogas is a valuable renewable energy source that forms naturally via microbial fermentation under anaerobic conditions such as are found in small lakes or flooded fields, and in the stomachs of ruminants. It can be produced in a more controlled manner by microbial digestion of organic material (agricultural waste, manure, municipal waste, sewage, food waste, etc.) in the absence of oxygen [1, 2]. The major components are methane (CH_4), carbon dioxide (CO_2), and humidity at saturation at digestion temperature, with traces of hydrogen sulfide (H_2S) and some other gases and vapors [3, 4]. The most common applications of biogas are for heating, combined heat and power generation, and as vehicle fuel. Other applications that have been studied or tested are injection into the natural gas grid and H_2 production for fuel cells.

A study of different utilizations of biogas reported that biogas upgrading to fuel quality gives the highest exportable energy with a medium range (10%) energy demand [5]. Upgrading was done with a membrane process. Sweden is today the leading nation when it comes to biogas as vehicle fuel, with a projected yearly consumption of 1 TWh in 2020, in comparison to 100 GWh in 2002 [6, 7].

To use biogas as vehicle fuel, the gas must be upgraded to fulfill certain required specifications. The corrosive components (water vapor and sulfur) present in biogas must be removed. CO_2, which is one of the major components in the biogas, needs to be separated from the biogas because it dilutes/lowers the heating value of the gas. This results in a reduced burning capacity, which affects the performance of the engine [8].

5.2 Carbon Hollow Fiber Production

The precursor for the carbon hollow fibers used in this study was prepared using regenerated cellulose acetate (CA) by the dry/wet phase inversion fiber-spinning process. The fibers were spun using a pilot-scale spinning set up delivered by Philos, Korea. Details of the spinning process are described elsewhere [9]. A dope consisting of CA mixed with N-methylpyrrolidone and polyvinylpyrrolidone was used to spin CA hollow fibers, which were deacetylated batch-wise with a mix solution of NaOH in a short-chain alcohol. Then the deacetylated, dried, and now mainly cellulosic hollow fibers were carbonized at 550 °C under N_2 flow (CO_2 gas was also tried for some production batches) in a tubular three-zone furnace. The carbonization protocol typically consisted of a heating rate of 1 °C/min with several dwells during the heating sequence (in the region of 250 to 300 °C) and a final 2 h dwell at 550 °C. The oven system had no active cooling system so the cooling rate could only be actively controlled at temperatures higher than 100 °C. The oven was tilted ca. 5° in the longitudinal direction of the working tube to assist in the drainage of tar/water produced during the carbonization. Details of development and optimization of carbonization processes can be found in [10] and Chapter 1.

A lot of fibers would necessarily have to be produced in order to serve a decent-sized biogas upgrading facility; some degree of "bad" fibers in the individual carbon bundles produced is unavoidable. The bad fibers may be subdivided into different classes:

- Broken fibers.
- Self-looping or severely curled fibers.
- Fibers with kinks and weak spots.
- Collapsed fibers (section of fiber with wall cave-in).
- Two or more fibers fused together from trapped tar buildup.

Rectifying the different faults varies from simply shaking out the majority of the broken fiber from the bundle to the laborious scrutiny of individual fibers with barely (if any) visible weak spots. The thoroughness needed in the quality control is to some extent dependent on the frequency of the occurrence of defects. The greatest challenge was in our experience due to weak spots (which may be very hard to spot) and to too-curly fibers; both of these faults have a tendency to induce fiber breakage during operation.

5.3 Carbon Membranes for Biogas Upgrading

In general, the composition and quality of the biogas varies a lot, and is obviously highly dependent on the source or substrate on which the bacteria are fed.

This means that each individual plant will require customizations/adaptations. On top of this, the proposed use of the upgraded biomethane may also put constraints on the upgrading process. Similarly, local or global legislations on methane emissions in the slip (rejected stream or permeate) may add to the complexity of the upgrading plant. To utilize the biogas as a fuel, several issues must be clarified in the initial stages of planning:

- A biogas upgrading pre-study should be performed, as described in the next section.
- The intended end use of the final biomethane must be identified.

5.3.1 Biogas Upgrading Pre-Study

A pre-study of the applicability of (carbon) membrane technology for the upgrading of biogas to biomethane ought to start with a detailed high-resolution, time-averaged analysis of the composition of the particular biogas generated. Several components in the "normal" biogas are of interest: nitrogen gas, oxygen gas, sulfur-containing components (including SO_x and H_2S), alcohols, acids, nitrogen oxides, water vapor (expected to be at saturation at fermentation temperature), and, of course, CO_2 and methane.

Detailed knowledge of these parameters will assist in choosing the appropriate amount and order of any necessary pretreatment steps needed to comply with all environmental legislation and methane utilization standards, and to ensure the carbon membrane has a sufficient lifetime.

5.3.2 Biomethane Utilization

Three typical uses are common:

- Dual fuel (typical methane purity 80–85 vol%).
- Vehicle fuel (typical methane purity ≥ 95 vol%).

- Gas-grid injection (typically, inert less than 7 vol%, with specific demand for O_2 level).

The actual specifications for these three qualities are highly dependent on region and need to be obtained and complied with for each case.

5.4 Case Study/Operation of a Biogas Upgrading Pilot Plant Based on Carbon Membranes

The now-closed Norwegian company MemfoACT AS attempted to produce carbon membranes on a semi-industrial scale. The author of this chapter was a member of the core team of that company. As part of the company's business development, a pilot-scale biogas upgrading facility was designed, built, and tested at a municipal renovation company in southern Norway. Their generated biogas came mainly from biological waste from households and food production factories. Typical compositions of the generated biogas are given in Table 5.1.

Because Norway does not have a biomethane standard for vehicles, the Swedish biomethane standard for vehicle fuel was adopted. Important values are summarized in Table 5.2 [6].

5.4.1 Process Description

Carbon membranes are more selective for CO_2 than CH_4; hence, in the biogas upgrading process the CO_2 from the feed passes through the membrane (low-pressure side or permeate) and CH_4 remains on the high-pressure side (retentate). This means that the retentate is the desired product.

A few economical questions arise instantaneously in the planning of a new facility, such as the expected membrane module cost (cost of membrane

TABLE 5.1

Average biogas compositions for the facility in the southern part of Norway.

Component	Household waste
Methane (CH_4)	64 ± 3 mole %
Carbon dioxide (CO_2)	30–35 mole %
Nitrogen (N_2)	<1 mole %
Oxygen (O_2)	ca. 0 mole %
Hydrogen sulfide (H_2S)	1000 ppm
Water (H_2O), 35 °C	ca. 32 g/Nm^3

TABLE 5.2

Summary of important values in the Swedish biomethane gas standard [6].

Component	Standard
CH_4 (vol%)	96–98
H_2O (mg/Nm^3)	<32
Dew point (°C)	–60 °C at 250 bar (g)
$CO_2 + O_2 + N_2$ (vol%)	<4
O_2 (vol%)	<1
H_2S (ppm)	<23

mounted in a suitable pressure vessel) and the cost of a compressor operating in the expected operational pressure ranges for the upgrading process. Additionally, can a thorough pretreatment of the gas ahead of the actual membrane separation lead to the expected increase in membrane lifetime and at what cost? In other words, should the applied gas standard requirements toward all trace compounds other than nitrogen be dealt with upstream of the membrane (where there is a much larger gas volume flow to process), or can it be done post membrane?

Generally, there exists a cost minimum, at a normally moderate pressure, that should be sought in the design of any membrane purification process where the retentate is the sought-after product. Recently, a paper from He et al. [11] attempted to generalize this in a combined economical and process simulation paper on biogas upgrading. For a large-scale facility, this approach is clearly important in minimizing upgrading cost (potentially maximizing profit in the sale of the biomethane). However, in this reported small-scale pilot plant, the focus was more on being able to demonstrate the concept and identify the bottlenecks in operation. (The upgrading facility referred to in this chapter was planned well ahead of the study from He et al.) As the carbon production process was at a semi-industrial scale, it was decided to use the highest possible pressure (21 bar), because this would yield the largest possible permeate flux and hence require production of the smallest membrane area. (Owing to limited fabrication capacity, this meant that the membranes would be produced more quickly.)

The process flowsheet of the upgrading process with essential components is shown in Figure 5.1. The principal plan for the process is to remove H_2S and bulk water from the biogas stream upstream of the membrane and compressor, then upgrade the biogas to biomethane at an intermediate pressure, followed by high-pressure compression to biomethane storage (typically ≥ 200 bar). The raw biogas was available at 1.03 bar from the digester and a blower was used to increase the pressure to 1.3 bar. This both reduces the physical size of the compressor and ensures stable flow to the upgrading system, which was situated several hundred meters away from the digester.

Carbon Membrane Technology

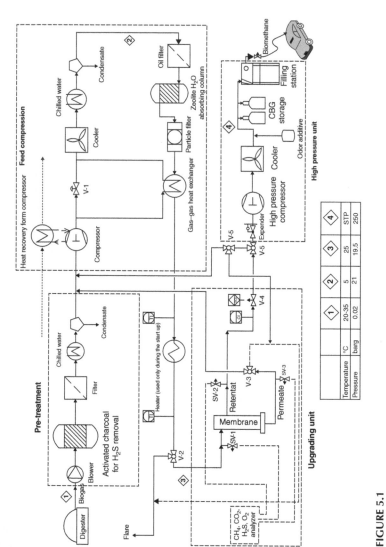

FIGURE 5.1

Process flowsheet of the biogas-upgrading pilot plant based on a carbon hollow fiber membrane [14].

An activated charcoal adsorber system was used to remove most of the H_2S and bring it down to 5 ppm (mole base) in the biogas stream. A filter was installed after the adsorber and before the water removal system to remove entrained particulates. At site, upstream from the upgrading unit, a microbial H_2S bioreactor was designed, built, and attempted to use but sadly was not operated successfully. The problem was twofold: there were difficulties in seeding the column with a thriving biofilm and also in controlling and keeping the pH within the desired operational window. The rationale behind using a H_2S bioreactor bulk remover was to attempt to spend less on charcoal (reduce running cost).

The bulk water was removed by means of a chiller (dew point: 4 °C at 1.3 bar) then the gas was reheated and led via the upgrading compressor (including a water- and oil-removal system) to a dual-tank zeolite water adsorption system where the fuel quality specs relating to water dewpoint/H_2O content (Table 5.2) were met. Although the carbon membranes are reported in the literature to be tolerant to water humidity levels up to 30% relative humidity (RH) [12], we chose to minimize the potential risk of membrane exposure to liquid water as a consequence of humidity condensation during unscheduled stops.

As shown in Figure 5.1, the membrane skid was next. This was a pilot plant, so the utilization and purity of the captured CO_2 in the permeate was not of any concern in its running. Consequently, a single-stage operation was applied. Regarding the operation of any membrane modules, a few parameters are user defined and can be tweaked to yield the desired product: (1) The ratio of permeate flow rate and feed flow rate is defined as stage cut (q_p/q_f). This means that retentate purity is a function of retentate flow rate and retentate pressure; these are controllable by the operator. (2) The method by which the driving force for separation is applied (i.e., the local chemical potential (partial pressure) difference for a given component between the feed-retentate side and the permeate side); that is, the way the module is connected is important in obtaining maximum membrane area utilization. Biomethane purity in the non-permeate stream also obviously depends on the CO_2/CH_4 selectivity of the carbon hollow fiber membrane. However, this selectivity is a material property and not something the operator can manipulate. Generally, carbon membranes possess high CO_2/CH_4 selectivity and can be operated at high pressure; thus, a sufficiently high pressure ratio can be achieved if required.

The biogas upgrading pilot plant, containing carbon hollow fiber membranes, was operated to achieve fuel-quality biomethane. The plant was designed to process 60 Nm^3/h of raw biogas at pressures up to 21 bar. The initial tests reported here were performed for a feed flow rate of 10 Nm^3/h at 21 bar pressure. Compressed and dried biogas, as described previously, entered an array of custom-designed cylindrical multi-modules (up to six). A photograph of one such module is shown in Figure 5.2. Each multi-module could contain up to 24 medium-sized carbon hollow fiber modules

(a)   (b)   (c)   (d)

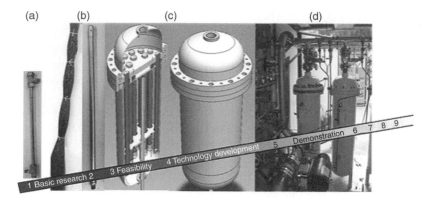

FIGURE 5.2
Technology readiness level according to the EU Commission/upscaling from lab to pilot scale;
(a) lab-scale module, (b) medium-sized module, (c) multimodule, (d) membrane pilot plant [14].

($\approx$0.5–2 m^2 each). The membrane feed gas temperature was regulated by an electric heater, and pressure/biomethane purity was controlled by a modulating valve (v-4 in Figure 5.1, also visible in the center of the left-hand-side photograph in Figure 5.3). The membrane pressure, temperature, and flow of the two outlet streams, permeate and retentate, were monitored. Online inline infrared analyzers were used to monitor the composition of the feed stream and the two membrane outlet streams. To ensure biomethane purity,

FIGURE 5.3
Photographs showing the biogas upgrading membrane plant.

an additional dedicated Fourier transform infrared CO_2 sensor was installed in the retentate. However, this makes the system "blind" to any nitrogen contamination or nitrogen dilution of the upgraded biomethane.

A portable infrared analyzer was also available and used to measure the composition of any streams that may not be routed through the online analyzers; this was particularly handy during troubleshooting of unscheduled stops. The permeate side was equipped with a vacuum pump, but this was only intended as a means of being able to commission the system with less membrane area installed (it was, however, sometimes helpful in identifying module problems during operation) – normal operation is without the vacuum pump engaged. High-pressure compression up to 250 bar and odor addition were performed before storage of the vehicle fuel. Photographs of this section of the plant are shown in Figure 5.4.

5.4.2 Membrane Modules Description

The designed multi-module system (MMS) comprised up to 24 medium-sized modules: each module is made up of up to 2000 carbon hollow fibers, which were individually tested for strength (this was done to reduce the probability of fiber breakage during construction of the individual modules), in bundles with effective areas ranging from 0.5 to 2 m^2. The outer diameter of the hollow fiber was in the range of 150–300 micron (micrometer, μm, 10^{-6} m) and its wall thickness was 30–50 micron. The feed was on the shell side (outside the fibers) of the module and the permeate flows internally (tube side) along the fibers. The assemblage, testing, and performance of each medium-sized module are reported elsewhere [13].

The MMS was designed to attempt to accomplish maximum efficiency of the membranes. Structural strength, low fouling tendency, membrane replacement, and ease of cleaning the MMS were important considerations for its

FIGURE 5.4
Photographs showing the biomethane compression and storage bottle bank.

application in a biogas upgrading plant. The MMS is 0.324 m in diameter and 1 m in active length and consists of three parts: (1) The vertical tank, which has both feed and retentate connecting ports and three legs with screws to secure it to the skid. (2) A middle part, to insert the medium-sized modules, consisting of two round plates with holes according to the outer diameter of the medium-sized modules. (This plate is needed to avoid short-circuiting the feed to the retentate, leaving virtually no flow over the membrane surface.) One partition plate on the top separates the permeate section from the feed section and another between feed and retentate helps to hold the modules firmly and avoid them bumping into each other. (3) The lid on the top with the permeate connection. The arrangement of the medium-sized modules inside the MMS is shown in Figure 5.2. After the assembly and before fitting the lid, each medium-sized module was again tested for any leakage (fiber breakage) using very low air pressure (slightly higher than atmospheric pressure) and soapy water. If any broken fibers were detected, they were marked/identified and it was attempted to seal the bore *in situ* using Loctite 3090 instant glue.

The MMS was then coupled with "diameter nominal" DN-flanges on the permeate side (leak-proofed by inserting gaskets in between). Compression fittings in national pipe threads were used on feed and retentate connections, then pressurized and filled with gas by adjusting the feed, retentate, and permeate valves manually in the initial stage. The valves were adjusted to fill in such a way that the fibers are treated gently on the surface and there is no excessive pressure difference between feed and retentate across the partition plate inside the MMS. This meant that it took approximately 1 to 3 h to pressurize the system.

5.4.3 Pilot Operation

Although there were a few start-up and tuning-in challenges for the upgrading system, as expected for a novel membrane type applied in a new setting, the system was successfully run for days as shown in Figures 5.5 and 5.6 (this part of the work has also been published previously [14]). As Figure 5.5 reveals, a big challenge in the initial runs was that the performance dropped severely in the course of hours after start-up. This was puzzling for quite a long time as laboratory testing of the modules, even when pressurized more quickly, did not reveal this tendency for delayed fiber breaking.

On close examination, the breakages were located just above or in the vicinity of one of the legs of the outer housing of the multi-module housing. As the individual modules are only tightly suspended to the housing in the top plate, the propagation of vibrations and a resonance phenomenon became the suspect. An external company was hired to investigate and their measurements revealed that vibration hotspots could indeed be detected. By implementing flexible tubing, to the modules on the feed-retentate side and on the suspended permeation piping, and vibration dampening of the

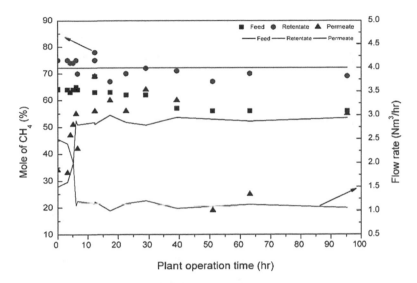

FIGURE 5.5
Results of one multi-module system, tested for a feed flow of 4 Nm³/h [14].

mounting plate for the modules, the problem was eliminated and a much more stable operation was experienced, as can be seen in Figure 5.6.

5.4.4 Challenges and Suggestions

Although the carbon membrane pilot plant successfully obtained vehicle fuel, challenges remained regarding both the membranes and the surrounding systems. We will focus here on issues related to the membrane or the module. The fibers, which were sorted manually, are randomly packed and mass transfer coefficients for random packings are smaller than those for regular dense packings. Flow through the randomly packed hollow fiber bundle can be highly nonuniform.

The effective membrane area was very much reduced because of selective and manual clogging (done with the intent of "saving" faulty modules). Furthermore, regions where fibers come in close contact may create sections of high pressure drop. The gas velocity through these regions is much lower than in regions where fiber spacing is larger, yielding higher mass transfer coefficient in these regions. On the other hand, in high velocity regions, there are increased chances of fiber breakage if any weak point occurs on the fiber surface. It may result in the formation of large flow-channels and hence a bypassing effect that results in process selectivity loss.

The MMS design for 24 medium-sized modules was not quite helpful: individual module housing instead of MMS housing would have made it easier to isolate and mend or replace the modules with bad performance.

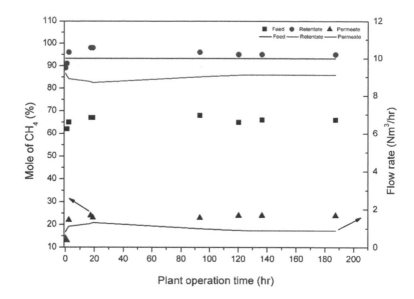

FIGURE 5.6
Carbon membrane separation process for biogas upgrading; flow rates are shown as solid lines and CH_4 contents as dots in the graph [14].

The process of dismantling the MMS to take out the medium-sized module, finding and clogging the broken fibers, and reassembling the MMS increased the probability of fiber breakage in neighboring modules inside the MMS and was very time-consuming. The shell-side feed configuration might have damaged the fibers due to high-pressure feed flow.

Toward the end of the pilot runs, an updated module housing was developed to house a single module (shell feeding) with a during-operation-removable permeate cap. This improved module design was only tested for a maximum of three modules in parallel on site. Using flexible metal tubes, this also allowed us to investigate to what extent, if any, gravity would influence the separation. The feed was fixed to the system but the retentate was controlled by a manual needle valve and the permeate simply vented over the roof. The module was simply rotated during operation (without having to disconnect any tubing) and the resulting permeate and retentate compositions measured with the handheld analyzer. As expected, the counter-current flow pattern was better than the co-current pattern, but to our surprise we actually obtained a significant (10 to 15%) performance increase by letting gravity pull the CO_2 out of the bore. In other words, having the permeate outlet pointing downwards was a significant improvement over the MMS design.

One should bear in mind that even though the module is potted at both ends the individual arrangement of the fibers inside the module cover would probably vary slightly with orientation; indeed, this is a reason not to design a system of horizontal modules. It was also found that, in vacuum permeate operation, having the feed on the bottom and the retentate on top (so gravity helps keep the heavy CO_2 from flowing to the retentate) gave approximately 5% better performance: the same module (equal membrane area) could produce a larger volume flow of biomethane with the same specifications.

The bore-side feed configuration could have been more efficient in the MMS system. The potential drawback of a bore-fed system is that the module must be opened at both ends. This will almost double the workload relating to potting, as both sides will need to be cut open, and the module housing would have to be redesigned. Based on our initial experience with a few fibers (Figure 5.2a), the bore-side feed seems to reduce the aging of the carbon if exposed to even higher than 30% RH humid gas relative to a comparable shell feeding. This was never thoroughly followed up.

The membrane production cost at a semi-industrial production plant was about $100/m^2$, but due to a decrease in membrane effective area the membrane cost doubled for the biogas pilot plant, which ultimately increased the total cost involved and production costs of any future upscaling of the plant. Because of this the site owner chose to build their full-scale plant based on water-scrubbing technology and the pilot is now sadly obsolete.

5.5 Conclusion

A complete and operational biogas upgrading system based on carbon membranes produced from regenerated cellulose has been designed and operated. A novel multimodule system containing 24 medium-sized modules was successfully installed and operated at 21 bar feed pressure to obtain vehicle fuel from biogas. The carbon hollow fiber membranes achieved 97 mol% CH_4 with up to 98% CH_4 recovery in a single-stage process. However, recovery was not a particular target for the customer. Pretreatment of biogas was performed prior to membrane separation, in order to meet the Swedish fuel quality standards. The pretreatment consisted of removing H_2S with a charcoal bed and H_2O by temperature swing followed by zeolite absorption. The plant was run successfully continuously for days (often 1 week nonstop) and the membranes used in this study yielded consistent results. It was observed that the shell-side feed configuration is not very efficient in the MMS because the fibers may damage or break with high-pressure feed flow. The required downtime for the facility to correct any module issues strongly suggest that, until the field of carbon membranes is more mature, a single pressure vessel per module is probably better.

References

[1] P. Weiland, Biogas production: current state and perspectives, *Appl. Microbiol. Biotechnol.*, 2010, 85(4), 849–860. doi: 10.1007/s00253-009-2246-7

[2] J.A. Lie, Synthesis, performance and regeneration of carbon membranes for biogas upgrading – A future energy carrier, Norwegian Uni. of Sci. and Techn., PhD Thesis, 2005.

[3] M. Hå, *Biogas för fordonsdrift - kvalitestsspecifikation, kommunikations forskningsberedningen (KFB4)*, Stockholm (in Swedish), 1997.

[4] K. Wågdahl, Distribution AV biogas I naturgasnätet, Svenskt Gastekniskt Center, Sweden, 1999. http://www.sgc.se/ckfinder/userfiles/files/SGC101.pdf.

[5] R. Rautenbach, K. Welsch, Treatment of landfill gas by gas permeation – Pilot plant results and comparison to alternatives, *J. Membr. Sci.*, 1994, 87(1), 107–118. https://doi.org/10.1016/0376-7388(93)E0091-Q.

[6] M. Persson, Utvärdering av uppgraderingstekniker för biogas, Lund/Malmö, 2003. http://www.sgc.se/ckfinder/userfiles/files/SGC142.pdf.

[7] J. Forsberg, Biogas Grid in Mälardalen Valley (in Swedish: Biogasnät i Mälardalen), Svenskt Gastekniskt Center AB, Malmö, 2014, 6. http://www.sgc.se/ckfinder/userfiles/files/SGC300web(1).pdf.

[8] S. Bari, Effect of carbon dioxide on the performance of biogas/diesel dual-fuel engine, *Renew. Energy*, 1996, 9(1), 1007–1010. https://doi.org/10.1016/0960-1481(96)88450-3.

[9] M.-B. Hagg. J. A. Lie, Carbon membranes, US20100162887 A1, 2010.

[10] S. Haider, J. A. Lie, A. Lindbråthen, M.-B. Hagg, Pilot–scale production of carbon hollow fiber membranes from regenerated cellulose precursor – Part I: Optimal conditions for precursor preparation. *Membranes*, 2018, 8(4), article number 97. https://doi.org/10.3390/membranes8040097.

[11] X. He, Y. Chu, A. Lindbråthen, M. Hillestad, M.-B. Hagg, Carbon molecular sieve membranes for biogas upgrading: Techno-economic feasibility analysis. *J. Cleaner Production*, 2018, 194, 584–593 http://dx.doi.org/10.1016/j.jclepro.2018.05.172.

[12] C. W. Jones, W. J. Koros, Characterization of ultramicroporous carbon membranes with humidified feeds, *Ind. Eng. Chem. Res.*, 1995, 34(1), 158–163. http://dx.doi.org/10.1021/ie00040a014.

[13] S. Haider, A. Lindbråthen, J. A. Lie, I. C. T. Andersen, M.-B. Hagg, CO_2 separation with carbon membranes in high pressure and elevated temperature applications. *Separ. Purif. Technol.*, 2017, 190, 177–189. http://dx.doi.org/10.1016/j.seppur.2017.08.038.

[14] S. Haider, A. Lindbråthen, J. A. Lie, P. V. Carstensen, T. Johannessen, M.-B. Hagg, Vehicle fuel from biogas with carbon membranes; a comparison between simulation predictions and actual field demonstration. *Green Energy Environ.* 2018, 3(3), 266–276. http://dx.doi.org/10.1016/j.gee.2018.03.003.

6

Carbon Membranes for Natural Gas Sweetening

Evangelos P. Favvas[a], Sotirios P. Kaldis[b] and Xuezhong He[c]

[a]*Membranes & Materials for Environmental Separations Laboratory, Institute of Nanoscience and Nanotechnology, NCSR*

[b]*Chemical Process & Energy Resources Institute, Centre for Research & Technology Hellas*

[c]*Department of Chemical Engineering/Faculty of Natural Sciences, Norwegian University of Science and Technology (NTNU)*

6.1 Introduction

The first studies of separation membranes began in the eighteenth century (1748) when the Abbé Jean-Antoine Nollet, a French experimental physicist, proved that the bubbling phenomenon in decompressed liquids was caused by dissolved air. In 1824, René-Joachim-Henri Dutrochet, a French physiologist, discovered the phenomenon of osmosis in natural membranes [1]. In 1855, Adolf Eugen Fick, a German physiologist, introduced Fick's law of diffusion, which describes the diffusion of a gas across a fluid membrane [2]. In 1861, Thomas Graham, a Scottish chemist, studied the diffusion of gases and laid the foundation of gas and vapor separation through polymeric membranes [3]. He could be called the father of modern dialysis, which is the usage that occupies the biggest share of the membrane market. In 1887, Jacobus Henricus van't Hoff, a Dutch physical and organic chemist, proposed the famous van't Hoff equation for osmotic pressure (π). This work was awarded the first Nobel Prize in Chemistry, in 1901 [4]. After these fundamental works, many other pioneering papers were published during the first half of the twentieth century; the "golden age" of membrane technology (1960–1980) began in 1960 with the invention by Loeb and Sourirajan [5] of the first asymmetric integrally skinned cellulose acetate reverse osmosis membrane.

Nowadays, together with water treatment/cleaning processes and hemodialysis processes, gas separation processes using membranes are a top priority of academic and industrial interest because natural gas (NG), which exists in deep underground reservoirs, usually contains several non-hydrocarbon components that must be separated and removed. Two of these are hydrogen sulfide (H_2S) and carbon dioxide (CO_2). This process is called "sweetening" [6] and is one of the major separations in the NG industry. The huge amounts of acid-pumped NG demand efficient solutions that are cheaper than existing methods. For example, the emirate of Abu Dhabi alone is expected to produce up to 6.8 billion cf/d of NG by 2020 [7]; the crude NG in this area contains more than 15% H_2S and sometimes up to 50% or even higher [8]. This is an unfortunate fact that many other producer countries also have to confront, which is why many research projects have been funded worldwide in an attempt to solve this major problem.

H_2S removal processes can be either physical-chemical or biological. If, for instance, the aim is to produce liquified NG or to remove N_2 cryogenic processing, then CO_2 must be reduced to less than 50 ppm in order to avoid solidification in exchangers, pipes or turbo expanders [9]. H_2S, in the presence of water, forms a weak, corrosive acid that causes premature failure of valves, pipelines and pressure vessels. CO_2 is also corrosive in the presence of water and lowers the heating value. When NG is used as domestic fuel, it becomes necessary to remove H_2S because of the health hazards associated with it. The threshold limit value for prolonged exposure of H_2S is 10 ppm [10].

Until now, three main technologies for NG sweetening have been applied [11, 12]:

1. **Absorption**
   - **Physical absorption**: Physical absorption processes are generally most efficient when the partial pressures of the acid gases are relatively high, because the partial pressure is the driving force for the absorption. Specifically, these processes are recommended for use when the partial pressure of the acid gas in the feed is greater than 50 psi yet the solvents are very sensitive to pressure [13].
   - **Chemisorption**: For chemisorption process the following reagents are typically used:
   - Hot potassium carbonate solution.
   - Amines. Alkanolamines, known as the Benfield process, are mostly used to absorb CO_2 and H_2S from the feed gas [14]. This is the most suitable solution in cases when the acid gas partial pressure is low and low levels of acid gas are desired in the residue gas stream. For CO_2 removal the following amines are extensively used:

- Monoethanolamine.
- Diethanolamine.
- Methyl diethanolamine.

The H_2S-rich concentrated acid gas is routed to a sulfur recovery unit to be converted into elemental sulfur by the well-known Claus process, which was first brought on-stream in 1973:

$$H_2S + \frac{3}{2}O_2 \rightarrow SO_2 + H_2O$$

$$2H_2S + SO_2 \rightarrow 2H_2O + 3S^0$$

$$H_2S + \frac{1}{2}O_2 \rightarrow S^0 + H_2O$$

2. **Adsorption**: Acid gases and water can be effectively removed by physical adsorption on synthetic zeolites. Applications are limited because water displaces acid gases on the adsorption bed [15].

   In these processes, the separation step entails the formation of molecular complexes that must be reversed through a significant increase in temperature. The heating and subsequent cooling of sorbents to prepare them for the next sorption cycle is thus highly energy consuming. In addition, diffusion limitations result in slow uptake and regeneration kinetics, rendering these systems inefficient to process gas with large amounts of H_2S and CO_2.

3. **Membranes**: Membranes can be used to remove bulk CO_2 and H_2S, preferably at high feed pressures. Different gases pass through the membrane and, depending on their different permeabilities (P), separation is achieved. The main mechanism of gas separation is therefore the difference in diffusion rate of each gas through the membrane, which depends on the sorption and/or molecular sieving effect [16]. In comparison with the other NG separation techniques, the membrane process offers a viable energy-saving alternative since it does not require any phase transformation.

6.2 Natural Gas

Energy consumption differs between developing and developed countries (Figure 6.1). Developed countries consume the majority of the energy, but even in developing countries the requested energy increases faster and faster

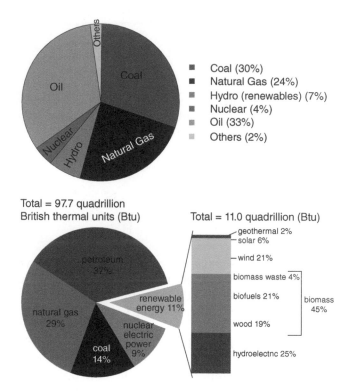

FIGURE 6.1

Top: World total primary energy consumption by fuel in 2015 [17], and Bottom: US energy consumption by energy source for 2017 [18].

from year to year. It must be noted that the absolute energy consumption is not proportional to each country's population. For example, although Americans make up less than 4.5% of the world's population they consume about 25% of the total global energy.

World energy resources can be divided into three types: (1) fossil fuel, (2) nuclear fuel and (3) renewable resources; fossil fuels are the most used energy sources worldwide and can be divided into the three subcategories coal, NG and oil. By nuclear fuel, we mainly mean uranium resources, which are estimated to be available until 2500 [19]. Renewable resources can be classified into the following main types: (1) wind power, (2) solar energy, (3) hydropower and (4) wave and tidal power, biomass and geothermal. Among all the above-described energy sources, oil is recorded as having the highest consumption, 33% (2015 data); after that is coal, 30%, and NG, 24% [17]. The annual global production of NG is estimated as 3680.4 billion cubic meters [20]. In the United States only, in 2017, about 950.000 million cubic meters of NG were produced, as shown in Table 6.1 [18].

TABLE 6.1

NG production (in million cubic meters) in the USA [18].

Gross withdrawals					
	2013	2014	2015	2016	2017
From gas wells	304.676	286.663	265.365	206.369	190.195
From oil wells	153.044	169.900	185.125	180.806	187.697
From coalbed wells	40.374	37.012	33.596	30.329	28.140
From shale gas wells	337.891	395.725	447.952	505.385	538.542
Total	**835.984**	**889.300**	**932.037**	**922.889**	**944.574**

The total NG produced in the United States increases year by year (see Figure 6.2) and at the same time cleaner fuels are necessary for both economic and environmental purposes.

As we can see from Table 6.1 the amount of NG produced from gas wells decreased while NG from shale gas wells increased. The existence of new NG reserves encourages the community and at the same time is a reason for developing new technologies for further purification of this highly acidic NG.

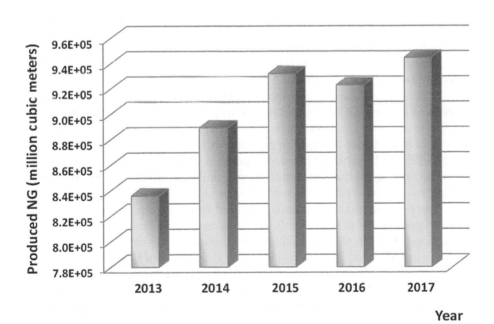

FIGURE 6.2
Natural gas production in the past 5 years in the USA.

NG forms deep beneath the earth's surface. A typical NG specific gravity is 0.59, with a gross heating value of 38.7 MJ/m^3 (dry basis) [21]. NG is used both as a fuel and for making chemicals and commercial materials.

NG consists mainly of methane but also contains small amounts of hydrocarbon gas liquids and non-hydrocarbon gases. Commercial NG (high CH_4 purity) has a higher heating value per unit mass (~50 MJ/kg, lower heating value) compared with other hydrocarbon fuels such as butane, gasoline and diesel fuel. Furthermore, it is recorded to be the fuel with the lowest carbon content per unit mass; during combustion it releases about 30% and 43% less CO_2 than oil and coal, respectively [22, 23]. A typical composition is about 93.9% methane, 4.2% ethane, 1% nitrogen, 0.5% carbon dioxide, 0.3% propane, 0.03% isobutane, 0.03% normal-pentane, 0.01% oxygen and some other elements in trace amounts. Although raw NG is mainly composed of methane (generally from 75% to 95%), it also contains unwanted components such as acid gas impurities. The major two acid gases are CO_2 and H_2S; both should be removed in order to prevent corrosion to pipeline, infrastructure and engines. CO_2 not only affects the calorific value but also in the presence of water forms a weak acid. It is a fact that in some cases the acid gases could be at very high levels, even more than half of the total amount of the raw produced NG.

The problem of high acid gas concentration in NG must be solved by numerous exploiter oil companies [24]. One of the most critical processes in the gas industry is the removal of acid gases, which is conventionally done by gas-liquid absorption-stripping processes. According to the US pipeline specifications, the NG criteria for sale are: maximum water content of 7 lbs/mmcf, maximum CO_2 content of 2% and maximum H_2S content of 4 ppm vol [23]. It is absolutely necessary that the acid gases and water traces should be removed from the NG before its use in industry. This increases the heating value of the NG, reduces corrosion, prevents atmospheric pollution and also avoids the water freezing in pipelines and accomplishes distribution of the NG [25]. In the current applied processes, mainly amine units and cryogenic technologies, the separation step entails the formation of molecular complexes that must be reversed through significant increase in temperature. The heating and subsequent cooling of sorbents to prepare them for the next sorption cycle is thus highly energy consuming. In addition, diffusion limitations result in slow uptake and regeneration kinetics, rendering these systems inefficient to process gas with large amounts of H_2S and CO_2.

Based on this anticipated capacity, there is a need for the development of new, economically efficient, technologies for NG purification. Membrane technology provides a promising solution for this application, especially in remote areas such as offshore platforms [26]. Thanks to the fact that membrane units are compact (they have a small footprint), reliable (no moving parts) and require less operation energy (no phase change is required to achieve separation) [27, 28], they are often reported as the solution for the future.

6.3 Membranes for Natural Gas Sweetening

During the past few years many types of membrane [29], such as metal membranes [30–32], polymeric membranes [33, 34], ceramic and zeolite membranes [35, 36] and mixed-matrix membranes [37–45], have been developed and their ability to separate specific gas molecules studied. Although a number of polymers exhibit permselective performance for gas mixtures, only a few glassy polymers are suitable for preparing asymmetric gas separation membranes and the most widely used are polysulfones (PSf), polyethersulfones (PESf), polyetherimides (PEI) and polyimides (PI). Polymeric membranes have also been evaluated in high-pressure separation applications, with discouraging results due to their inherent disadvantages related to poor temperature resistance [46], corrosion and swelling phenomena [47, 48]. Zeolite membranes usually have polycrystalline structures through which molecules can permeate. The molecules can pass through both the zeolite pores (intracrystalline pathways) and the gaps between crystals (intercrystalline pathways) [49, 50]. The existence of these gaps severely degrades the gas-separation efficiency of zeolite membranes [51].

Polymeric membranes can separate H_2S and CO_2 from NG [52, 53] in low temperature streams mainly composed of CH_4, C_2H_6, and N_2. Until today there have been limited reports referring to gas sweetening processes using membrane technology. These works are focused on different kinds of materials with different gas solubility and permeability properties. Currently, the only commercially viable membranes for CO_2 removal are based on polymeric materials such as cellulose acetate, polyimides, polyamides, polysulfones, polycarbonates, and polyetherimide. In particular, polyimides combine excellent thermal properties and chemical stability and can easily form numerous membrane structures. In addition, they exhibit a wide range of CO_2 permeabilities and show good potential structural variations. Polyimide membranes were initially used for hydrogen recovery, while they have also been modified for CO_2 removal applications. For instance, the crosslinked poly(ethylene glycol) acrylate contains branches with methoxy end groups and exhibits high CO_2 permeability (570 barrer) and high CO_2/CH_4 selectivity (12) at 35 °C, which can be further increased by decreasing temperature [54]. One approach to remove CO_2 from NG streams has been the design of polymeric membranes by different crosslinking methodologies using the poly(ethylene oxide)-based family of materials. In this way the membranes exhibit good solubility-selectivity performance for CO_2/CH_4 separation relative to conventional size-selective membranes, especially in strongly plasticizing gas streams. Their advantages could be further enhanced by operating at low temperatures, which may provide additional flexibility for integrating membranes into NG processes. These materials could also be used, potentially, to remove polar molecules such as H_2S, H_2O and NH_3 from mixtures with light gases such as CH_4, N_2 and H_2, given the fact that the ethylene oxide units exhibit good affinity towards these small polar gas molecules [16].

Furthermore, poly(ether urethanes) are reported as good materials for H_2S/CH_4 separation. Specifically, separation factors of about 20 for CO_2/CH_4 and about 100 for H_2S/CH_4 are reported for a ternary $CH_4/CO_2/H_2S$ mixture at 20 °C [55]. Studies by Mazur and Chan [56] describe the removal of ~95% CO_2 and ~90% H_2S, respectively, using cellulose acetate membranes. Schell and Houston [57] obtained a selectivity of 20 to 30 for CO_2 over CH_4 and 75 to 110 for H_2S over C_3H_8 using cellulose acetate membranes.

Currently, membrane technology research focuses on thin film, composite polymer membranes, and associated process configurations to increase the flux of CO_2, thereby reducing the required membrane area (U). This system, called the MTR Polaris membrane, shows a permeance ten times higher than cellulose acetate (the most common material used for CO_2/CH_4 separation during NG processing) with the same CO_2/N_2 selectivity [58]. The MTR Polaris membrane has been demonstrated in modules of 20–50 m^2 of membrane area in the field of NG processing [59].

In general, glassy polymers (cellulose acetate, polyimides etc.) provide high separation factors for CO_2 over methane (about 10–20) whereas rubber polymers (ether-amide block copolymers etc.) are better candidates for H_2S/CH_4 separation $(20 < \alpha_{H_2S/CH_4} < 40)$ [26]. On the other hand, inorganic membranes can separate CO_2 present in small concentrations in a hot (200–600 °C) gaseous stream mainly composed of CO and H_2 (syngas) or in large concentrations (10%–12%) in a flue gas stream (N_2: 78%–80%, O_2: 2%–6%, [CO, NOx, SO_2]: 10^2 ppm), based mainly on two mechanisms: (a) molecular sieving, i.e. exclusion by size of molecules larger than CO_2 and (b) preferential sorption of the CO_2 and blocking the flow through the membrane of less-sorbable gases such as nitrogen [60].

6.4 Polymeric Membranes and Ionic Liquids for Natural Gas Sweetening

Although polymeric membranes provide numerous advantages over inorganic membranes, some significant properties make inorganic membranes (carbon and zeolite) more attractive for specific applications. Specifically, the polymeric membranes have the main advantages of low production cost and easily handling but, at the same time, provide poor chemical and thermal resistance. On the other hand, carbon membranes provide excellent chemical and thermal stability and transcend the trade-off between permeability and selectivity. However, they are characterized as brittle materials and present higher production costs [61].

According to Baudot [62] only three main polymeric membrane types have been tested at both a laboratory and a pilot scale for H_2S removal from NG:

1. *Membranes based on rubbery polymer selective layers*

   Here, selectivity is achieved based on the affinity difference between the components to be separated. The affinity is mainly defined by the difference in sorption coefficient. This kind of membrane generally presents high and very high permeabilities for low molecular weight components. The most common families of this type of polymeric membrane material are: (1) non-polar rubbery polymers [63], (2) non-polar rubbery polymers impregnated with polar solvents [64, 65] and (3) polar rubbery polymers [66].

   Each category takes advance of the different diffusion of H_2S and CO_2 versus CH_4 through the polymer matrix or the impregnated solvent.

2. *Membranes with a selective layer made of block copolymers*

   Membrane Technology and Research (MTR Inc.) developed and supply commercial spiral-wound membranes with a selective layer made of numerous polyether block amides. This membrane is well known with the commercial trademark of Pebax®, owned by Arkema. According to Amo et al. [67], the associations of Pebax®-based membranes with a sulfatreat process led to significantly higher operation costs (from 20% to 40%) over the amine process alone for flow gases containing a few percent of H_2S and lower than 140,000 Nm^3/h, whereas a Pebax®-membrane/amine coupling was more valuable than a stand-alone amine process for gas flows containing more than 5% H_2S and higher than 23,000 Nm^3/h.

3. *Membranes with a selective layer made of glassy polymers*

   This kind of polymeric membrane is mainly suggested for CO_2 removal applications included in the NG sweetening process. In this rigid type of polymer (polyimides, polysulfones, cellulose acetate etc.) the separation mainly takes place as a result of the difference in the kinetic diameter of the component. In the case of H_2S/CH_4 the selectivity is low because the kinetic diameters of those gases are rather similar [68].

The development of new polymeric membrane materials for testing in the NG sweetening process remains attractive both scientifically and economically. Owing to high levels of interest from academia and industry, many membranes for NG sweetening have been proposed in the recent literature. Table 6.2 presents some important results from the relevant literature concerning the use of polymeric membranes for acid gas removal from NG.

During recent years, attention has been given also to other types of candidate materials for acid gas removal from NG, including a large number of different ionic liquids (ILs) [81]. ILs are salts in the liquid state. They are known as "solvents of the future" as well as "designer solvents." Some ILs with noteworthy properties concerning their ability to separate CO_2 and/or H_2S from CH_4 have recently been described. The reported ideal selectivity values from ILs such as Bmim-BF_4, Emim-CH_3SO_4, HOemim-NO_3, Pmg-L, 2mHEAPr, [C_4mim][CH_3SO_3] etc. are between 13 and 22 for CO_2/CH_4 and between 10 and 75 for H_2S/CH_4 [82–85].

6.5 Inorganic and Polymeric Membranes: A Comparative Approach

Nowadays, commercial polymeric membrane systems are widely used in numerous separation processes. In fact, this is the membrane type that has defined the membrane market worldwide. As mentioned previously, this is thanks to the fact that many different types of low-cost polymeric materials are available, which can be processed into membrane structures that provide good permeability and selectivity factors in industrial separations. At the same time, this kind of polymeric membrane material is not appropriate in "extreme" environments, such as high temperatures, corrosive environments etc.

Inorganic membranes, however, can be used in a wide variety of environments. They can achieve difficult separations, such as propane–propylene [86] and C_4–hydrocarbons [87] and there are many voices saying that it will not be long before inorganic membranes begin to replace polymeric membranes. For both polymeric and inorganic membranes, the crucial point is to achieve the optimum combination of permeability and selectivity factors. The higher the permeability value, the smaller is the membrane area required, and therefore the capital cost of the system is lowered. Inorganic membranes have the potential to exceed the upper bound of the Robeson's plot [88] and this is stable proof of their potential to be commercialized in the future. In this context, ultra-microporous (0.3–0.5 nm) membranes such as zeolite and carbon membranes (and carbon molecular sieve membranes) have shown promise [89].

Table 6.3 summarizes some important differences between polymeric and carbon membranes [61].

TABLE 6.2

Polymeric membranes for acid gas removal from natural gas.

Polymer	P (CO$_2$)	P(H$_2$S)	P(CH$_4$)	α CO$_2$/CH$_4$	α H$_2$S/CH$_4$	Feed (bar)	Gas composition (CH$_4$/CO$_2$/H$_2$S)%	Ref./Year
Matrimid® 5218	18000			10.6	9.5	26.53	(balance/4/800 ppm)	[69]/1995
PTMSP, poly(1-trimethylsilyl-1-propyne)				0.85		1.38	10.5% CO$_2$/46% CO/1.5% H$_2$S	[70]/2006
Poly(ester urethane urea)							91.6/5.4/3	[71]/2008
Supported ionic liquid membranes (PVDF + BMIM BF$_4$)	30–180	160–1100		25–45	130–260	—	PG(iii)	[72]/2009
Seragel (butadiene–sulfone block copolymer)					4.6	1	3.91% H$_2$S/96% CH$_4$	[73]/2010
GENERON	45	4	1.3	34.6	3	—	PG(iii)	[74]/2011
PVTMS	1600	350	220	7.3	1.59	—	PG(iii)	
SILAR	2000	1195	545	3.67	2.19	—	PG(iii)	
Fluorinated, 6FDA-based polyamide–imide							PG(iii)	[75]/2012
Polybenzimidazole (PBI)/stainless steel composite membrane	0.15(i)	0.005(i)				3.34	55% H$_2$/41% CO$_2$/1% CO/1% CH$_4$/1% N$_2$/1% H$_2$S	[76]/2012
Poly(amide-6-b-ethylene oxide) (PEBA1657)	89	126				3	PG(iii)	[53]/2012
Cellulose acetate (pure)	8.66	8.71		29.5	29.7	34.5	60% CH$_4$/20% CO$_2$/20% H$_2$S	[77]/2013
PDMS-coated PES	116(i)	44	29		10.6	10	97.5/2.1/0.4	[78]/2014
6FDA polyimide/6F-PAI-1/Pebax	100	487		10	49	4.5	PG(iii)	[79]/2014
Polysulfone (PSF)	16.6(ii)	52.3(ii)		12.8	40.5	10	PG(iii)	[80]/2014

(i) Permeance in GPU, (ii) permeance in 10^{-6}, cm^3/cm^2 s cm Hg, and (iii) PG: pure gas measurements.

TABLE 6.3

Polymeric membranes versus carbon membranes.

	Polymeric membrane	Carbon membrane
Separation mechanism	Solution diffusion	Knudsen diffusion: >10 Å
		Surface diffusion: <50 Å
		Capillary condensation: >30Å
		Molecular sieving: <6 Å
Advantages	Low production cost	Excellent chemical stability
		Surpasses the trade-off between permeability and selectivity
		Excellent thermal stability
		Can be used at aggressive operation
Disadvantages	Poor thermal resistance	Vulnerable to adverse effects from exposure to organic contaminants and water vapor
	Poor chemical resistance	Brittle
	Arduous to reach the trade-off between permeability and selectivity	High production cost

Although the inorganic membranes seem to be the optimum type of membrane material for separations, the fact is that there are few scientific works mentioning these membranes as candidate materials for NG sweetening processes. This is mainly because the inorganic membranes are extremely expensive (ceramic, zeolitic materials) or difficult to construct as large modules (carbons).

In the market to date, only polymeric membrane modules are available for CO_2 removal from NG. A previous work by He reported that there are six main commercially available membrane modules for NG sweetening (Table 6.4) [90].

6.6 From Polymeric Material to Carbon Membranes

Nanoporous carbon membranes, and especially microporous hollow fiber membranes, have recently been identified as up-and-coming materials for gas separation applications (Figure 6.3).

TABLE 6.4

Membranes for NG sweetening [90].

Membrane	Material	Company	Module
Separex™	Cellulose acetate	UOP	Spiral wound
Cynara®	Cellulose acetate	NATCO	Hollow fiber
Prism®	Polysulfone	Air products	Hollow fiber
Cytop	Perfluoropolymers	MTR	—
Medal	Polyimide	Air liquid	Hollow fiber

In general, compared with other inorganic materials, carbon membranes are more insensitive, durable when exposed to organic solvents or vapors, present high resistance to oxidation, adsorb various gases well and are highly hydrophobic. The most important thing is that it's possible to tailor the pore size in carbon membranes by selecting the nature of the precursor and varying its pyrolysis parameters [91, 92]. Moreover, the use of special techniques leads to a uniform pore size distribution. Therefore, these kind of membranes are possible candidates for applications such as the purification of H_2, the separation of olefin/paraffin mixtures, CO_2 recovery from NG, the enrichment of syngas, water treatment and the production of oxygen enriched from air [93, 94].

Pyrolysis, or carbonization, is the process in which a suitable polymer, the precursor material, is heated in a controlled environment (inert, vacuum or oxidative, reductive) up to a high temperature according to an accurate heating protocol (heating rate, thermal soak time, environment, gas flow etc.). In Figure 6.4 a block diagram describes the four main "transition" steps,

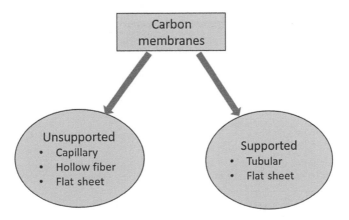

FIGURE 6.3
The main configurations of carbon membranes.

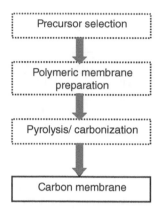

FIGURE 6.4
From polymeric material selection to carbon membrane preparation.

from polymeric material selection to carbon membrane preparation. Figure 6.5 represents a typical tube pyrolysis oven able to apply both a gas and a vacuum carbonization environment. During recent decades a wide range of polymeric precursors have been used for the preparation of carbon membranes. In order to successfully transform from the polymeric phase to this carbon material, some basic polymeric material property criteria must be satisfied. Some of these are: (1) the polymeric material must have a high aromatic carbon content, (2) the glass transition temperature, T_g, should be high,

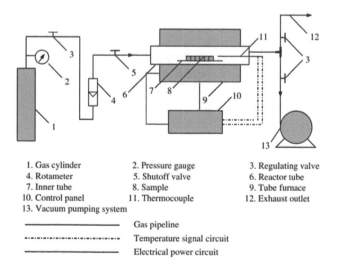

FIGURE 6.5
Tubular pyrolysis oven system [95].

(3) the polymer should be chemically stable and (4) the separation properties should be characterized between "satisfactory" and "superior" [61, 96]. As described and discussed in the next section, the most important reported families of polymeric precursor materials are: (1) the polyimides [97–99], (2) the polyetherimides [100, 101] and (3) the phenolic resins [102, 103].

Although these three polymer families are the most appropriate for use as carbon membrane precursors, many other polymers such as polyphenylene oxide [104, 105], polyacrylonitrile [106], phenolic resin [107, 108] etc. have also been used quite successfully.

The idea that the pyrolysis conditions have the most important influence on the developed membrane's structure and its gas permeability/selectivity properties is strongly questionable. Specifically, both pore dimension and microstructure characteristics strongly depend on the nature of the precursor polymer. The chemical structure of the polymeric precursor material can define the pore population and the shape of the prepared carbon matrix [109, 110].

Overall, carbon hollow fiber membranes are more expensive than polymeric derivative membranes due to the necessity of one more process, which requires high energy. This process is pyrolysis, for which a high-temperature furnace must be used for many hours. However, the derivative membranes have the advantages of often having better selectivity and permeability coefficients as well as higher chemical and of course thermal stability [111].

6.7 Carbon Membranes for Natural Gas Sweetening

As previously mentioned, carbon membranes can be produced by carbonization/pyrolysis of a variety of carbon-containing materials, mostly under vacuum, inert or oxidative atmosphere [112]. Recently, numerous synthetic precursors have been used to form carbon membranes, such as polyimide and its derivatives, polyacrylonitrile, phenolic resin, polyfurfuryl alcohol, polyvinylidene chloride–acrylate terpolymer, phenol, formaldehyde and cellulose [45]. Among them, the polyimides are one of the most promising candidates for carbon membrane production for CO_2 separation [113–115]. The developmental membranes, produced by pyrolysis of polyimide precursors, are characterized by stable nanoporous structures that exhibit satisfactory permeability and selectivity properties [91, 116–119].

In order to be applicable to an industrial process the membrane must possess two properties: high selectivity and high production capacity. To satisfy these criteria, both the membrane structure and the membrane module must have a high permeability and selectivity performance. An asymmetric membrane structure and the hollow fiber configuration of the membrane module are the two characteristics that provide the best combination of these two targets. Asymmetric membranes can be produced via a phase-inversion process, which can be adapted to produce membranes in the form of thin tubes or fibers.

An important advantage of hollow fiber membranes is that compact modules with a very high membrane surface area can be formed. However, this advantage is offset by the generally lower fluxes of hollow fiber membranes compared with flat-sheet membranes made from the same material [26]. Nonetheless, the development of hollow fiber membranes by Mahon and the group at Dow Chemical in 1966 [120] and their later commercialization by Dow, Monsanto, Du Pont and others, represents a major event in membrane technology. Of all possible membrane configurations, hollow fibers are the most promising for gas separation applications.

The properties of derivative carbon materials strongly correlate with the structural and selectivity properties of precursor polymeric membranes. To this end a good choice of selected precursor is the first step. For any candidate carbon membrane polymeric material, there are four major properties that need to be satisfied: (1) a high glass transmission temperature, T_g, (2) high aromatic content, (3) chemical stability and (4) high separation properties [61].

Especially in the case of the NG sweetening, the carbon membranes require high mechanical and chemical stability, since the operating pressure is high and the composition of the NG can create oxidative conditions. These requirements are the main reason that carbon membranes are not yet commercially available for NG sweetening processes, but also only a small number of scientific papers are available in this field.

Table 6.5 summarizes some of the major polymer types that have been used as precursor materials for carbon membranes and that are potentially applicable in NG sweetening processes, emphasizing the polyimide materials that are the most-studied precursor materials for producing carbon membranes for CO_2 separations. Numerous carbon membranes, both in flat sheet and hollow fiber configurations, have been reported as candidate materials for CO_2 separation/removal from NG streams. However, no carbon membrane module able to be installed in real industrial CO_2 separation plants is yet available, and this is a major priority of the industry, with many projects focused on this target worldwide.

H_2S is an extremely toxic component of NG that corrodes pipes and engines and must be removed from the NG prior to use. As shown in Table 6.5, carbon membranes are promising for CO_2 removal, but few data have been reported for H_2S removal from NG using carbon membranes. Current technologies for H_2S removal include absorption, adsorption, conversion of H_2S into elemental sulfur and a membrane reactor for H_2S decomposition and desulfurization. In addition, a hollow fiber membrane contactor has been in the limelight due to its potential to overcome problems such as foaming and

TABLE 6.5

Carbon membranes for acid gas (CO_2) removal from NG.

Precursor material	CO_2 permeability (barrer)	CO_2/CH_4 selectivity	Heat-treated (°C)	Membr. Config.±	Ref./Year
Polyimides (PI)					
Various PI	88[i]	190	500 and 550	HF	[121]/1994
Matrimid® and Kapton® polyimides	12, 92	33, 16	475–700	FS	[122]/ 1999
BTDA-TDI/MDI (P84) copolyimide	0.5[i]	51	900	HF	[123]/2000
6FDA/BPDA-DAM	80[i]	80	800	HF	[124]/2003
6FDA/BPDA-DAM and Matrimid 5218	30 and 13	73 and 83	800	HF	[125]/2002
BTDA-TDI/MDI (P84) copolyimide	1808, 738, 499	22, 37, 89	550, 650 and 800	FS	[111]/2004
Matrimid® and P84 polyimides	191, 611	169, 61	800	FS	[97]/2004
P84 polyimide	1.25[i]	51.4	800	HF	[126]/2008
PIM-6FDA-OH	5040, 556	38, 93	630 and 800	FS	[127]/2013
Polyimide of intrinsic microporosity (PIM-6FDA-OH)	557 and 471	15 and 59	440 and 800	FS	[128]/2013
Polyetherimide (PEI)					
Polyetherimide	0.15[ii]	155	600	FS	[129]/1999
PEI/polyvinylpyrrolidone (PVP)	1.66[i]	55	700	HF	[130]/2012
PEI/polyvinylpyrrolidone (PVP)	0.69[i]	69	800	HF	[131]/2012
Polyetherimide	0.183[iii]	6.5	600	FS	[132]/2013
Other (PSF/resins)					
Phenolic resin	24.5[iv]	160	800	FS	[133]/1999
Poly(furfurylalcohol) (PFFA) resin	0.035[ii]	36.8	850	FS	[134]/1998
PSF-beta/matrimid	12.6[i]	150	800	HF	[135]/2007
Poly(benzimidazole) (PBI)	0.16	88.88	800	FS	[136]/2009
PBI/matrimid	36.6	131.65	800	FS	[136]/2009
Cellulose acetate	60	110	650	HF	[137]/2009
Cellulose (pilot scale report)	0.11–0.045[v]	104.2–66.7	—	HF	[138]/2018
Cellulose	200		650	HF	[139] /2019

FS: flat sheet, HF: hollow fiber, (i) permeance in GPU, (ii) ($cm^3/cm^2 \cdot psi, min$), (iii) $m^3/m+\cdot bar \cdot h$, (iv) (($mol/m^2 \cdot s \cdot Pa$) $\times 10^{-10}$), (v) m^3 (STP)/($m^2 \cdot h \cdot bar$).

loading [140]. Another technique for H_2S treatment is microbiological treatment of the H_2S-containing gases [141]. In membrane technology the H_2S is removed by two main methods: (1) by using the membrane as a barrier between the liquid and the gas phase where it allows the transport of H_2S from the gas phase to the liquid phase. In this technology polymeric materials, such as polysulfones, polypropylene, polyvinylidene fluoride [142, 143] etc., are used; and (2) by using catalytic membrane reactors.

The thermal decomposition of H_2S into hydrogen and sulfur takes place in the catalytic membrane reactor. Since H_2S decomposition at high temperatures is attributed to the required high energy and endothermicity of the reaction, the use of a membrane reactor to selectively move the reaction products from the reaction zone in order to shift the equilibrium reaction forward sets the conditions for the application of a lower process temperature [140]. In this process glass and inorganic membranes have been used, but not carbon membranes. Specifically, membranes made of Vycor glass [144], alumina, microporous zirconia–silica [145], amorphous silica [146], metallic (Pd) [147] and composite metal containing Pt [148] have been used in catalytic H_2S decomposition reactions. Furthermore, typical H_2S desulfurization into elemental sulfur and water via catalytic oxidation is moving from fixed-bed reactor technology to catalytic membrane technology. Carbon membranes with the requested membrane properties, and porous membranes (α-Al_2O_3) impregnated with catalyst, have also been tested in this kind of process [149].

6.8 Natural Gas Sweetening Process Simulation and Optimization

The objective of membrane system modeling is to design membrane process systems with optimal conditions and configurations in order to accomplish optimal purities at minimum capital and operational costs. There have been numerous efforts over past decades. The first pioneering works presented by Weller and Steiner [150, 151] describe a method to calculate the permeate composition of a binary gas mixture in a cross-flow membrane under negligible pressure drop. Brubaker and Kammermeyer [152] were the first to deal with multicomponent membrane separation in a perfect mixed membrane. Shindo et al., 1985 and Pan, 1986 examined the separation of multicomponent gas mixtures with various flow patterns (one-side mixing, perfect mixing, cross flow, countercurrent and co-current) [153, 154]. One key assumption is the introduction of the Hagen–Poiseuille approximation for describing the permeate pressure drop inside the hollow fiber. The basic conclusion is that a countercurrent flow pattern results in higher permeate purity.

Coker et al. [155] considered the hollow fiber membrane as a series of stages in the axial direction and developed a model taking into account pressure-dependent permeability coefficients and bore-side pressure gradients. They concluded that the existence of a boundary layer at the membrane interface (known as concentration polarization) results in the formation of pressure and concentration gradients and reduces the driving force. Scholz et al. [156] developed and validated a model taking into account concentration polarization, the Joule–Thomson effect, pressure losses and real gas behavior. They concluded that the non-ideality of a CO_2 gas mixture is significant and could affect the modeling results.

Kaldis et al. [157] and Khalilpour et al. [158] investigated the effect of various design and operating parameters such as area, length, pressure, stage cut and selectivity on the separation of CO_2 from flue gas over co/countercurrent hollow fiber membranes. They showed that when the concentration of CO_2 in the inlet feed is rather low (less than 15%), membranes cannot accomplish high permeate purity and/or high recovery. A very useful conclusion is that high CO_2 selectivities (i.e. higher than 50) are less advantageous than expected.

6.9 Process Design of Natural Gas Sweetening Plants

Raw NG varies substantially in composition, depending on the source reservoir. Methane is always the major component, typically 75%–90% of the total, but NG also contains undesirable impurities such as water, carbon dioxide, nitrogen and hydrogen sulfide that need to be removed before it can be stored as compressed gas or delivered to the pipelines. CO_2 content varies from 3% to 10% and needs to be removed in order to avoid pipeline corrosion and minimize atmospheric pollution [159]. Owing to the very low concentration of CO_2, a single-stage membrane, as shown by modeling, cannot produce high purity permeate or residue even at very high inlet pressures and/or large membrane areas. The solution is a combination of a multiple membrane stages, in parallel or in series. Such combinations, however, result in higher capital costs, due to high membrane area, and operating expenses, due to high compression costs. In such cases membranes are less competitive than conventional gas separation technologies. The viability of membrane systems will be dependent on process synthesis, configuration and design [26].

The cascade model (shown in Figure 6.6) is the most general multistage membrane design. The downstream membranes enrich the desired components in the permeate and upstream membranes strip remaining traces of desired components. In the case of NG sweetening, an upstream stripping section is required in order to minimize the concentration of sour gas

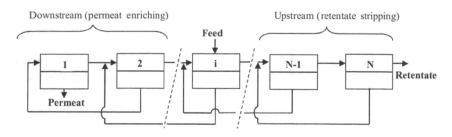

FIGURE 6.6
Schematic of cascade membrane system with recycles [160].

(CO_2 and H_2S) in retentate. Furthermore, in order to minimize NG loss in the permeate stream, a downstream stripping section is also needed.

Pettersen and Lien [161] and Avgidou et al. [162] studied the behavior of several single-stage and multi-stage permeator systems. It is proven that at very low feed concentrations or with low-efficiency membranes, a two- or three-stage membrane system is the most technoeconomically optimal configuration. Similar results were reported by Carapellucci and Milazzo, who highlighted in 2003 that the two-stage design is the best option for enriching the CO_2 stream [163]. Ho et al. [164] investigated cascade systems with the introduction of vacuum instead of pressurized permeation. According to their results, the vacuum two-stage system with retentate recycling could achieve the highest CO_2 purity. A slight modification examined by Merkel et al. [165] also introduced a two-stage and two-step membrane system considering countercurrent flow at vacuum mode and using air as a sweep gas to generate driving force. Alshehri et al. [160] introduced a multi-stage membrane network superstructure of possible flowsheet configurations. An optimization formulation was then developed and solved using an objective function that minimizes the costs associated with operating and capital expenses. Their main conclusion is that developing membranes with higher permeance rather than high selectivity will help membrane technology's competitiveness.

The membrane NG sweetening systems are either single or two-stage processes [166]. The single-stage design is suitable for remote and unmanned facilities, where maintenance and troubleshooting are more difficult. Furthermore, the absence of a compression station in the single-stage design diminishes the additional capital and operating cost. However, due to membranes' non-perfect selectivity the single-stage design is less advantageous than conventional technologies (e.g., amines). It is therefore necessary that a two-stage process (single recovery stage) be introduced. With this scheme, the overall separation outcome and economical indexes are more attractive than those of conventional technologies [166].

Finally, several studies have recommended the use of hybrid units consisting of a membrane unit followed by an amine unit [167]. The hybrid units can accomplish targeted separation at a lower cost but such units are not widely utilized in industry and they have increased operational and trouble-shooting complications.

6.10 Natural Gas Sweetening Process: A Financial Approach

There have been many technoeconomic studies evaluating membranes for gas separation applications and the removal of CO_2. In an early pioneering work by Vandersluijs et al. [168], technical and economic aspects, as well as the mitigation costs of polymer membranes for the recovery of CO_2 from flue gases of a power plant, were investigated. For 50% product purity and 75% recovery for CO_2, the minimum achievable cost was estimated to be US$48 per ton of CO_2 avoided. They showed that in order for membranes to become economically competitive with absorption technologies, a CO_2/N_2 selectivity higher than 200, along with high permeability, would be required.

Kazama et al. [169] carried out an economic analysis to evaluate a polyimide hollow fiber membrane with high CO_2 permeance and a CO_2/N_2 selectivity of 40. They showed that membrane systems become economically advantageous over amine-based capture systems if the inlet CO_2 concentration is higher than 25%. Almost the same conclusions were drawn by Bounaceur et al., who analyzed a single-stage membrane separation in comparison with amine-based solvent processes [170]. Amine absorption is shown to be the best option when a high CO_2 purity is desired, due to the large energy penalty for membrane systems. However, when the CO_2 concentration of flue gas is more than 20%, membrane separation is more advantageous. This study also concluded that for membranes with CO_2/N_2 selectivities less than 50, the membrane process is not feasible: selectivities higher than 100 are required.

Ho et al. [171] compared the feasibility of a single-stage hollow fiber membrane process with three membranes with CO_2/N_2 selectivities of 20, 30 and 50, respectively, resulting in a total sequestration cost of 55–61 US$/ton CO_2 avoided. They concluded that at such membrane selectivities, a high feed gas compression rate is required, which makes membranes economically non-competitive against monoethanolamine systems. The same conclusion was obtained by Favre, 2007, comparing dense polymeric membrane processes with amine absorption for post combustion [172]. Ho et al. [164] compared the two major strategies for producing the driving force for membrane gas separation at coal-fired power plants: compressing the feed gas stream versus vacuum pumping on the permeate side. They showed that although the vacuum strategy required a relatively large membrane area, it achieved 35% less capture cost per ton of CO_2 avoided than the compression design. The conclusion that a single-stage membrane process is not cost-competitive with amine solvents was also drawn by Zhao et al. who investigated the membrane properties as well as various operating conditions [173].

Merkel et al. [165] developed a two-stage membrane system using combustion air as sweep gas that carries a fraction of permeated CO_2 back, thus increasing the CO_2 concentration of the flue gas. Their study claimed that the membrane process with combustion air sweep for 90% CO_2 capture would account for 16% of the power output of a coal-fired power plant and result in a capture cost as low as \$23/ton CO_2. Their study showed that improving the CO_2 permeance may be more important than increasing CO_2/N_2 selectivity and with this design configuration the total capture cost may be diminished to \$23/ton CO_2.

The two-stage membrane configuration was also examined by Hussain and Hägg, 2010, using a CO_2-selective facilitated transport membrane with a selectivity of 200 [174]. The technoeconomic comparison with amine absorption showed that, with that specific configuration, a membrane process is feasible, even for low CO_2 concentration (10%) flue gas, and achieves more than 90% CO_2 purity and recovery.

In a thorough analysis, Zhai and Rubin evaluated the performance and costs of single- and multi-stage membrane configurations [175]. It was shown that multi-stage membrane systems are able to achieve a 90% removal efficiency and 95% or more product purity for CO_2. The economic assessment results indicate that the multi-stage membrane system may be capable of achieving the separation targets at a cost that is 15% less than that of an amine-based capture process.

Ahmad et al. [176] used a new two-dimensional cross-flow model in Aspen HYSYS for the simulation of membrane systems for CO_2 removal from NG. In their paper, different configurations including single-stage (with and without recycling) and multiple-stage membrane systems (with permeate and retentate recycling) were investigated. Their results indicate that the gas-processing cost is lowest for a double-stage membrane system with permeate recycling, making this the optimal design for the membrane separation system.

The same authors in a later study examined the temperature and pressure dependence of membrane permeance for CO_2 separation from methane [177]. The temperature drop due to Joule–Thomson cooling and its effect on membrane permeance of both CO_2 and CH_4 has been studied. It was shown that highly non-ideal conditions (e.g., higher CO_2 concentration in the feed) strongly affect the product quality, methane loss, stage cut, compressor power and gas-processing cost of hollow fiber separation systems.

Peters et al. [178] studied both amine sweetening and membrane acid gas removal and estimated the total capital investment of each system. Each system was studied and optimized separately to achieve 2% CO_2 in sweet gas and above 90% CO_2 in acid gas. They showed that total capital investment for membranes was lower than that for amines. Lock et al. [179] studied the effect and trend of the recycling of CO_2 on the NG-processing cost under different operating conditions. The general trend is that increasing the recycling increases the membrane area and compressor power while improving hydrocarbon recovery. Thus, a trade-off must be determined among these parameters in all cases to minimize the gas-processing cost.

He et al. [180] conducted process simulation and optimization to evaluate process feasibility of novel fixed-site-carrier (FSC) membranes. The membranes were used for high pressure NG sweetening, and experimental data from high pressure small pilot-scale module tests were obtained. Two-stage membrane systems were designed to evaluate process feasibility for CO_2 removal from different NG sources. The simulation results indicated that membrane systems could be a potential candidate for CO_2 removal from NG.

However, before bringing the membranes into a large-scale demonstration or commercial application the membrane performance needs to be further improved, especially at high pressure operation; influences of water vapor on FSC membrane performance should be examined and membrane durability should be tested with a real NG.

Carbon membranes, despite being more selective and more robust than polymer membranes, have not drawn the attention of researchers until recently. Recently, Haider et al. [181, 182] compared, from a technoeconomic point of view, the use of polymeric and carbon-based membranes in biogas upgrading systems. The cost of the carbon membranes was five times that of polymeric membranes, although they needed two stages to achieve the targets in terms of recovery and purity, instead of the three stages needed by polymer membranes. Energy analysis showed that the use of carbon membranes may reduce electricity consumption by up to 22%. A two-stage carbon molecular sieve membrane system with different CO_2 concentrations in the feed was investigated by Chu and He [183]. By using HYSYS simulation, they found that the cost for processing a specific NG was significantly affected by the membrane performance, especially CO_2/CH_4 selectivity. The second-stage permeate pressure had a great influence on the cost when the feed content had higher CO_2.

6.11 Conclusions

The differences in gas separation and process models (multicomponent versus binary, multi-stage versus single-stage etc.) in existing membrane studies are considered to be one of the key sources of discrepancies among the technoeconomic studies. Furthermore, the costing scope and methodology appear inconsistent across existing studies: application of a common costing methodology could improve the clarity and consistency of cost estimates.

The most commonly used single-stage membrane configuration is not feasible for NG due to the low CO_2 content. If one stage is selected, the operating costs are much smaller with membrane technology but with the addition of a second stage, the operating costs become higher than other technologies. In most of the feed streams, at least two stages are required to meet the targets in terms of separation efficiency, even for membranes with high permselectivities.

A plant-level analysis is needed to help explore tradeoffs in meeting performance and cost objectives and identify the most promising system designs and R&D targets in terms of material properties, for improving membrane capture technologies.

Acknowledgments

E.P. Favvas and X. He would like to thank the Research Council of Norway for the partial funding of this work through the Petromaks2 program in the CO2Hing project (#267615). E.P. Favvas would also like to thank the project MIS 5002567, implemented under the "Action for the Strategic Development on the Research and Technological Sector", funded by the Operational Program "Competitiveness, Entrepreneurship and Innovation" (NSRF 2014-2020) and co-financed by Greece and the European Union (European Regional Development Fund) for the support of part of this work.

List of Acronyms

FSC	Fixed site carrier
GPU	Gas permeance unit (barrer without thickness correction) (1GPU = 3.35×10^{-10} mol m^{-2} s^{-1} Pa^{-1})
IL	Ionic liquid
NG	Natural gas
PBI	Polybenzimidazole
PES	Polyethersulfone
PFFA	Poly(furfurylalcohol)
PI	Polyimide

List of Symbols

P	Permeability, barrer (1 Barrer = 0.33×10^{-15} mol m m^{-2} s^{-1} Pa^{-1})
T	Temperature, K

T_g Glass transition temperature, K
U   Membrane area
α Selectivity

References

[1] L. Dongfei, Dual-layer asymmetric hollow fiber for gas separation, PhD Thesis, National University of Singapore, 2004.

[2] A. Fick, On Liquid Diffusion, *Phil. Mag.* 10 (1855) 30.

[3] T. Graham, On the absorption and dialytic separation of gases by colloid septa, *Phil. Mag.* 32 (1866) 401.

[4] J.H. van't Hoff, Osmotic pressure and chemical equilibrium, Nobel Prize in Chemistry, 1901.

[5] Loeb and S. Sourirajan, *Advan. Chem. Ser.* 38 (1962) 117.

[6] R.K. Abdulrahman, I.M. Sebastine, Natural gas sweetening process simulation and optimization: A case study of Khurmala field in Iraqi Kurdistan region, *J. Nat. Gas Sci. Eng.* 14 (2013) 116–120.

[7] https://www.bp.com/content/dam/bp/business-sites/en/global/corporate/pdfs/energy-economics/statistical-review/bp-stats-review-2020-full-report.pdf, accessed June 2020.

[8] T. Matsumoto, Abu Dhabi energy policy-energy problems plaguing Abu Dhabi and their implications for Japan, IEEJ (August 2013), http://eneken.ieej.or.jp/data/5115.pdf.

[9] A.J. Kidney, W.L. Parrish, *Fundamentals of Natural Gas Processing*, Taylor & Francis, Boca Raton, FL, 2006, pp. 93–127.

[10] *Occupational Safety & Health Administration*, www.osha.gov.

[11] J.S. Gudmundsson, *Natural gas sweetening & effect of declining pressure*, TPG4140 Project report, Norwegian University of Science & Technology, Trondheim, Norway.

[12] S.A. Ebenezer and J.S. Gudmunsson, Removal of carbon dioxide from natural gas for LPG production, in: carbon dioxide removal processes, 21 August 2011, Available from: http://www.ipt.ntnu.no/~jsg/studenter/prosjekt/Salako2005.pdf.

[13] *Engineering Data Book*, 12th ed., Volume 2, Sec. 21, *Hydrocarbon Treating*, Gas Processors Suppliers Association, Tulsa, OK, 2004, pp. 6–14.

[14] J. Tobin, P. Shambaugh et al., The crucial link between natural gas production and its transportation to market, in *Stages in the Production of Pipeline-Quality Natural Gas and NGLs*, 2006.

[15] *Engineering Data Book*, 12th ed., Volume 1, Sec. 2, *Product specification*, Gas Processors Suppliers Association, Tulsa OK, 2004, p. 3.

[16] H. Lin, E. Van Wagner, R. Raharjo, B.D. Freeman, and I. Roman, High-performance polymer membranes for natural-gas sweetening, *Adv. Mater.* 18 (2006) 39–44.

[17] *BP Statistical Review of World Energy*, June 2016, www.bp.com.

[18] *Natural Gas Annual, US Energy Information Administration*, https://www.eia. gov/naturalgas/annual/.

[19] *International Atomic Energy Agency, Global Ur Resources to Meet Projected Demand*: Latest Edition of "Red Book" Predicts Consistent Supply up to 2025, June 2, 2006, Vienna International Centre, www.iaea.org.

[20] *BP Statistical review of world energy*, June 2018, www.bp.com.

[21] *Union Gas Limited*, A Canadian natural gas storage, transmission and distribution company, https://www.uniongas.com/.

[22] L. Langston and S.L. Lee, A bright natural gas future, *Mech. Eng.* 132 (2010) 49.

[23] G. George, N. Bhoria, S. AlHallaq, A. Abdala, V. Mittal, Polymer membranes for acid gas removal from natural gas, *Separ. Purif. Technol.* 158 (2016) 333–356.

[24] P. Nederlof, N. Kaczorowski, D. Lawrence, The origin of H_2S in the Arab Reservoirs in Abu Dhabi, *Abu Dhabi Intern. Petroleum Exhib. & Conf.*, November 7–10, 2016, Abu Dhabi, UAE, SPE-183336-MS.

[25] S.M. Javaid Zaidi, Removal of acid gases from natural gas streams by membrane technology, *Proceedings of the 2nd Annual Gas Processing Symposium*, Qatar, January 10–14, 2010, pp. 139–144.

[26] R.W. Baker, *Membrane Technology and Applications*, 2nd ed., Wiley, New York, 2004.

[27] Z. Lai, G. Bonilla, I. Diaz, J. G. Nery, K. Sujaoti, M. A. Amat, E. Kokkoli, O. Terasaki, R. W. Thompson, M. Tsapatsis, D. G. Vlachos, Microstructural optimization of a zeolite membrane for organic vapor separation, *Science* 300 (2003) 456–460.

[28] M.P. de Haas, F. Choffat, W. Caseri, P. Smith, J.M. Warman, Charge mobility in the room-temperature liquid-crystalline semiconductor poly(di-nbutylstannane), *Adv. Mater.* 18 (2006) 44–47.

[29] E.D. Bernardo and G. Golemme, Membrane gas separation: A review/state of the art, *Ind. Eng. Chem. Res.* 48 (2009) 4638–4663.

[30] G. Alefeld and J. Völkl (editors), *Hydrogen in Metals I: Basic Properties*, Springer-Verlag, Germany, 1978.

[31] P. Liu, X. Ge, R. Wang, H. Ma, Y. Ding, Facile fabrication of ultrathin Pt overlayers onto nanoporous metal membranes via repeated Cu UPD and in situ redox replacement reaction, *Langmuir* 25 (2009) 561–567.

[32] N.A. Luechinger, S.G. Walt, W.J. Stark, Printable nanoporous silver membranes, *Chem. Mater.* 22 (2010) 4980–4986.

[33] U. Merten, P.K. Gantzel, Method and apparatus for gas separation by diffusion. US Patent 3415038, 1968.

[34] A. Hasimi, A. Stavropoulou, K.G. Papadokostaki, M. Sanopoulou, Transport of water in polyvinyl alcohol films: Effect of thermal treatment and chemical crosslinking, *Eur. Polym. J.* 44 (2008) 4098–4107.

[35] J. Caro, M. Noack, P. Kolsch, R. Schäfer, Zeolite membranes – State of their development and perspective, *Micropor. Mesopor. Mater.* 38 (2000) 3–24.

[36] J. Brinker, C.-Y. Tsai, Y. Lu, Inorganic dual-layer microporous supported membranes. US Patent 6536604, 2003.

[37] E.K. Chatzidaki, E.P. Favvas, S.K. Papageorgiou, N.K. Kanellopoulos, N.V. Theophilou, New polyimide–polyaniline hollow fibers: Synthesis, characterization and behavior in gas separation, *Eur. Polym. J.* 43 (2007) 5010–5016.

[38] H.J.C. Te Hennepe, D. Bargeman, M.H.V. Mulder, C.A. Smolders, Zeolite-filled silicone rubber membranes: Part 1. Membrane preparation and pervaporation results, *J. Membr. Sci.* 35 (1987) 39–55.

[39] J.-M. Duval, B. Folkers, M.H.V. Mulder, G. Desgrandchamps, C.A. Smolders, Adsorbent filled membranes for gas separation. Part 1. Improvement of the gas separation properties of polymeric membranes by incorporation of microporous adsorbents, *J. Membr. Sci.* 80 (1993) 189–198.

[40] M.-D. Jia, K.V. Peinemann, R.-D. Behling, Preparation and characterization of thin-film zeolite–PDMS composite membranes, *J. Membr. Sci.* 73 (1992) 119–128.

[41] S.F. Nitodas, E.P. Favvas, G.E. Romanos, M.A. Papadopoulou, A.Ch. Mitropoulos, N.K. Kanellopoulos, Synthesis and characterization of hydrogen selective silica-based membranes, *J. Porous Mater.* 15 (2008) 551–557.

[42] R. Mahajan, W.J. Koros, Factors controlling successful formation of mixed-matrix gas separation materials, *Ind. Eng. Chem. Res.* 39 (2000) 2692–2696.

[43] R. Mahajan, W.J. Koros, Mixed matrix membrane materials with glassy polymers. Part 1, *Polym. Eng. Sci.* 42 (2002) 1420–1431.

[44] R. Mahajan, W.J. Koros, Mixed matrix membrane materials with glassy polymers. Part 2, *Polym. Eng. Sci,* 42 (2002) 1432–1441.

[45] S. Chengwen, W. Tonghua, W. Xiuyue, Q. Jieshan, C. Yiming, Preparation and gas separation properties of poly(furfuryl alcohol)-based C/CMS composite membranes, *Separ. Purif. Technol.* 58 (2008) 412–418.

[46] J. E. Mark, *Polymer Data Handbook,* Oxford University Press, New York, 1999.

[47] J.D. Wind, S.M. Sirard, D.R. Paul, P.F. Green, K.P. Johnston, W.J. Koros, Carbon dioxide-induced plasticization of polyimide membranes: Pseudo-equilibrium relationships of diffusion, sorption, and swelling, *Macromolecules* 36 (2003) 6433–6441.

[48] J.D. Wind, S.M. Sirard, D.R. Paul, P.F. Green, K.P. Johnston, W.J. Koros, Relaxation dynamics of CO_2 diffusion, sorption, and polymer swelling for plasticized polyimide membranes, *Macromolecules* 36 (2003) 6442–6448.

[49] S.N. Wijenayake, N.P. Panapitiya, S.H. Versteeg, C.N. Nguyen, S. Goel, K.J. Balkus Jr., I.H. Musselman, J.P. Ferraris, Surface cross-linking of ZIF-8/polyimide mixed matrix membranes (MMMs) for gas separation, *Ind. Eng. Chem. Res.* 52 (2013) 6991–7001.

[50] Y.C. Hudiono, T.K. Carlisle, A.L. LaFrate, D.L. Gin, R.D. Noble, Novel mixed matrix membranes based on polymerizable room-temperature ionic liquids and SAPO-34 particles to improve CO2 separation, *J. Membr. Sci.* 370 (2011) 141–148.

[51] K. Varoon, X. Zhang, B. Elyassi, D.D. Brewer, M. Gettel, S. Kumar, J.A. Lee, S. Maheshwari, A. Mittal, C-Y. Sung, M. Cococcioni, L.F. Francis, A.V. McCormick, K.A. Mkhoyan, M. Tsapatsis, Dispersible Exfoliated Zeolite Nanosheets and Their Application as a Selective Membrane, *Science* 334 (2011) 72–75.

[52] S. Sridhar, B. Smitha, T.M. Aminabhavi, Separation of carbon dioxide from natural gas mixtures through polymeric membranes—A review, *Separ. Purif. Rev.* 36 (2007) 113–174.

[53] X. Ren, J. Ren, M. Deng, Poly(amide-6-b-ethylene oxide) membranes for sour gas separation, *Separ. Purif. Technol.* 89 (2012) 1–8.

[54] H. Lin, B.D. Freeman, Materials selection guidelines for membranes that remove CO_2 from gas mixtures, *J. Mol. Struct.* 739 (2005) 57.

[55] G. Chatterjee, A.A. Houde, S.A. Stern, Poly(ether urethane) and poly(ether urethane urea) membranes with high H2S/CH4 selectivity, *J. Membr. Sci.* 135 (1997) 99–106.

[56] W.H. Mazur, M.C. Chan, Membranes for natural gas sweetening and CO_2 enrichment, *Chem. Eng. Prog.* 78 (1982) 38–43.

[57] W.J. Schell, C.D. Houston, Industrial gas separations, *ACS Symp. Ser.* 223 (1982) 125–132.

[58] A. Brunettia, F. Scuraa, G. Barbieria, E. Drioli, Membrane technologies for CO_2 separation, *J. Membr. Sci.* 359 (2010) 115–125.

[59] S. Shelley, Capturing CO_2: membrane systems move forward, *Chem. Engin. Progress* 105 (2009) 42.

[60] N.K. Kanellopoulos (ed.), Recent advances in gas separation by microporous membranes, *Elsevier*, August 2000.

[61] W.N.W. Salleh, A.F. Ismail, T. Matsuura, M.S. Abdullah, Precursor selection and process conditions in the preparation of carbon membrane for gas separation: A review, *Separ. Purif. Rev.* 40 (2011) 261–311.

[62] A. Baudot, Natural gas sweetening, E. Drioli, L. Giorno (eds.), *Encyclopedia of Membranes*, Springer-Verlag, Berlin, Heidelberg, 2012, doi: 10.1007/978-3-642-40872-4_405-1.

[63] T.C. Merkel, R.P. Gupta, B.S. Turk, B.D. Freeman, Mixed-gas permeation of syngas components in poly(dimethylsiloxane) and poly(1-trimethylsilyl-1-propyne) at elevated temperatures. *J. Membr. Sci.* 191 (2001) 85–94.

[64] S. Kulprathipanja, Separation of gases from non-polar gases, US Patent 4,606,740 (1985).

[65] S. Kulprathipanja, S.S. Sudhir, Separation of polar gases from non-polar gases. US Patent 4,608,860 (1985).

[66] C.J. Orme, F.F. Stewart, Mixed gas hydrogen sulfide permeability and separation using supported polyphosphazene membranes, *J. Membr. Sci.* 253 (2005) 243–249.

[67] K. Amo, R.W. Baker, V.D. Helm, T. Hofman, K. Lokhandwala, I. Pinnau, M. Ringer, T.T. Su, L. Toy, J.G. Wijmans, Low-quality gas sulfur removal/recovery, DOE final report (1998) DE-AC21-92MC28133-01.

[68] J. Hao, P.A. Rice, S.A. Stern, Upgrading low-quality natural gas with H_2S- and CO_2-selective polymer membranes–Part II. Process design, economics, and sensitivity study of membrane stages with recycle streams. *J. Membr. Sci.* 320 (2008) 108–122.

[69] R.W. Baker, K.A. Lokhandwala, Sour gas treatment process including membrane and non-membrane treatment steps, United States Patent 5407466 (1995).

[70] T.C. Merkel, L.G. Toy, Comparison of hydrogen sulfide transport properties in fluorinated and nonfluorinated polymers, *Macromolecules* 39 (2006) 7591–7600.

[71] T. Mohammadi, M.T. Moghadam, M. Saeidi, M. Mahdyarfar, Acid gas permeation behavior through poly (ester urethane urea) membrane, *Ind. Eng. Chem. Res.* 47 (2008) 7361–7367.

[72] Y.-I. Park, B.-S. Kim, Y.-H. Byun, S.-H. Lee, E.-W. Lee, J.-M. Lee, Preparation of supported ionic liquid membranes (SILMs) for the removal of acidic gases from crude natural gas, *Desalination* 236 (2009) 342–348.

[73] M. Saeidi, M.T. Moghadam, M. Mahdyarfar, T. Mohammadi, Gas permeation properties of Seragel membrane, *Asia-Pac. J. Chem. Eng.* 5 (2010) 324–329.

[74] O.V. Malykh, A.Yu. Golub, V.V. Teplyakov, Polymeric membrane materials: New aspects of empirical approaches to prediction of gas permeability parameters in relation to permanent gases, linear lower hydrocarbons and some toxic gases, *Adv. Colloid Interface Sci.* 164 (2011) 89–99.

[75] J.T. Vaughn, W.J. Koros, J.R. Johnson, O. Karvan, Effect of thermal annealing on a novel polyamide–imide polymer membrane for aggressive acid gas separations, *J. Membr. Sci.* 401–402 (2012) 163–174.

[76] K.A. Berchtold, R.P. Singh, J.S. Young, K.W. Dudeck, Polybenzimidazole composite membranes for high temperature synthesis gas separations, *J. Membr. Sci.* 415 (2012) 265–270.

[77] C.S. K. Achoundong, N. Bhuwania, S.K. Burgess, O. Karvan, J.R. Johnson, W.J. Koros, Silane modification of cellulose acetate dense films as materials for acid gas removal, *Macromolecules* 46 (2013) 5584–5594.

[78] S. Saedi, S.S. Madaeni, A.A. Shamsabadi, PDMS coated asymmetric PES membrane for natural gas sweetening: Effect of preparation and operating parameters on performance, *Can. J. Chem. Eng.* 92 (2014) 892–904.

[79] J.T. Vaughn, W.J. Koros, Analysis of feed stream acid gas concentration effects on the transport properties and separation performance of polymeric membranes for natural gas sweetening: a comparison between a glassy and rubbery polymer, *J. Membr. Sci.* 465 (2014) 107–116.

[80] S. Sridhar, B. Smitha, S. Mayor, B. Prathab, T.M. Aminabhavi, Gas permeation properties of polyamide membrane prepared by interfacial polymerization, *J. Mater. Sci.* 42 (2007) 9392–9401.

[81] F. Karadas, M. Atilhan, S. Aparicio, Review on the use of ionic liquids (ILs) as alternative fluids for CO_2 capture and natural gas sweetening, *Energy Fuels* 24 (2010) 5817–5828.

[82] S. Mortazavi-Manesh, M.A. Satyro, R.A. Marriott, Screening ionic liquids as candidates for separation of acid gases - Solubility of hydrogen sulfide, methane and ethane, *AIChE J.* 59 (2013) 2993–3005.

[83] P.J. Carvalho, J.A.P. Coutinho, The polarity effect upon the methane solubility in ionic liquids: a contribution for the design of ionic liquids for enhanced CO_2/ CH_4 and H_2S/CH_4 selectivities, *Energy Environ. Sci.* 4 (2011) 4614.

[84] L. Crowhurst, P.R. Mawdsley, J.M. Perez-Arlandis, P.A. Salter, T. Welton, Solvent–solute interactions in ionic liquids, *Phys. Chem. Chem. Phys.* 5 (2003) 2790–2794.

[85] H. Salari, A.R. Harifi-Mood, M.R. Elahifard, M.R. Gholami, Solvatochromic probes absorbance behavior in mixtures of 2-hydroxy ethylammonium formate with methanol, ethylene glycol and glycerol, *J. Solution Chem.* 39 (2010) 1509–1519.

[86] K.A. Stoitsas, A. Gotzias, E.S. Kikkinides, Th.A. Steriotis, N.K. Kanellopoulos, M. Stoukides, V.T. Zaspalis, Porous ceramic membranes for propane–propylene separation via the p-complexation mechanism: unsupported systems, *Micropor. Mesopor. Mater.* 78 (2005) 235–243.

[87] M. Gehre, Z. Guo, G. Rothenberg, S. Tanase, Sustainable separations of C4-hydrocarbons by using microporous materials, sustainable separations of C4-hydrocarbons by using microporous materials, *ChemSusChem* 10 (2017) 3947–3963.

[88] L.M. Robeson, The upper bound revisited, *J. Membr. Sci.* 320 (2008) 390–400.

[89] T.A. Centeno, A.B. Fuertes, Carbon molecular sieve membranes derived from a phenolic resin supported on porous ceramic tubes. *Separ. Purif. Technol.* 25 (2001) 379–384.

[90] X. He, Membranes for natural gas sweetening, *Encyclopedia of Membranes*, E. Drioli and L. Giorno, eds., Springer, Berlin, Heidelberg, 2016.

[91] S.M. Saufi, A. F. Ismail, Fabrication of carbon membranes for gas separation – A review, *Carbon* 42 (2004) 241–259.

[92] A.F. Ismail, L.I.B. David, A review on the latest development of carbon membranes for gas separation, *J. Membr. Sci.* 193 (2001) 1–18.

[93] H.P. Hsieh, Inorganic membranes for separation and reaction, in H.P. Hsieh, *Membrane Science and Technology Series*, Vol. 3, Elsevier, Amsterdam, 1996.

[94] W.J. Koros, R. Mahajan, Pushing the limits on possibilities for large scale gas separation: which strategies? *J. Membr. Sci.* 175 (2000) 181–196.

[95] A.C. Lua, J. Su, Effects of carbonisation on pore evolution and gas permeation properties of carbon membranes from Kapton polyimide, *Carbon* 44 (2006) 2964–2972.

[96] J.N. Barsema, N.F.A. van der Veg, G.H. Koops, M. Wessling, Carbon molecular sieve membranes prepared from porous fiber precursor. *J. Membr. Sci.* 205 (2002) 239–246.

[97] P.S. Tin, T.S. Chung, A.J. Hill, Advanced fabrication of carbon molecular sieve membranes by nonsolvent pretreatment of precursor polymers. *Ind. Eng. Chem. Res.* 43 (2004) 6476–6483.

[98] E.P. Favvas, G.E. Romanos, F.K. Katsaros, K.L. Stefanopoulos, S.K. Papageorgiou, A.Ch. Mitropoulos, N.K. Kanellopoulos, Gas permeance properties of asymmetric carbon hollow fiber membranes at high feed pressures, *J. Nat. Gas Sci. Eng.* 31 (2016) 842–851.

[99] N. Sazali, W.N.W. Salleh, A.F. Ismail, K.C. Wong, Y. Iwamoto, Exploiting pyrolysis protocols on BTDA-TDI/MDI (P84) polyimide/nanocrystalline cellulose carbon membrane for gas separations, *J. Appl. Polym. Sci.* 136 (2019) 46901.

[100] M.G. Sedigh, M. Jahangiri, P.K.T. Liu, M. Sahimi, T.T. Tsotsis, Structural characterization of polyetherimide-based carbon molecular sieve membranes. *AIChE J.* 46 (2000) 2245–2255.

[101] B. Zhanga, Y. Wu, Y. Lu, T. Wang, X. Jian, J. Qiu, Preparation and characterization of carbon and carbon/zeolitemembranes from ODPA–ODA type polyetherimide, *J. Membr. Sci.* 474 (2015) 114–121.

[102] T.A. Centeno, J.L. Vilas, A.B. Fuertes, Effects of phenolic resin pyrolysis conditions on carbon membrane performance for gas separation, *J. Membr. Sci.* 228 (2004) 45–54.

[103] S.N.A. Jalil, D.K. Wang, C. Yacou, J. Motuzas, S. Smart, J.C. Diniz da Costa, Vacuum-assisted tailoring of pore structures of phenolic resin derived carbon membranes, *J. Membr. Sci.* 525 (2017) 240–248.

[104] M. Yoshimune, I. Fujiwara, K. Haraya, Carbon molecular sieve membranes derived from trimethylsilyl substituted poly(phenylene oxide) for gas separation, *Carbon* 45 (2007) 553–560.

[105] M. Yoshimune, K. Mizoguchi, K. Haraya, Alcohol dehydration by pervaporation using a carbon hollow fiber membrane derived from sulfonated poly(phenylene oxide), *J. Membr. Sci.* 425 (2013) 149–155.

[106] H. Fan, F. Ran, X. Zhang, H. Song, W. Jing, K. Shen, L. Kong, L. Kang, A hierarchical porous carbon membrane from polyacrylonitrile/polyvinylpyrrolidone blending membranes: Preparation, characterization and electrochemical capacitive performance, *J. Energy Chem.* 23 (2014) 684–693.

[107] S.C. Rodrigues, R. Whitley, A. Mendes, Preparation and characterization of carbon molecular sieve membranes based on resorcinol–formaldehyde resin, *J. Membr. Sci.* 459 (2014) 207–216.

[108] S. Roy, R. Das, M.K. Gagrai, S. Sarkar, Preparation of carbon molecular sieve membrane derived from phenolic resin over macroporous clay-alumina based support for hydrogen separation, *J. Porous Mater.* 23 (2016) 1653–1662.

[109] P.S. Tin, Y. Xiao, T.S. Chung, Polyimide-carbonized membranes for gas separation: structural, composition, and morphological control of precursors. *Separ. Purif. Rev.* 35 (2006) 285–318.

[110] B.T. Low, T.S. Chung, Carbon molecular sieve membranes derived from pseudo-interpenetrating polymer networks for gas separation and carbon capture, *Carbon* 49 (2011) 2104–2112.

[111] P.S. Tin, T.-S. Chung, Y. Liu, R. Wang, Separation of CO_2/CH_4 through carbon molecular sieve membranes derived from P84 polyimide. *Carbon* 42 (2004) 3123–3131.

[112] P. Pratibha, R. S. Chauhan, Membranes for gas separation, *Progr. Polym. Sci.* 26 (2001) 853–893.

[113] C. Staudt-Bickel, W.J. Koros, Olefin/paraffin gas separations with 6FDA-based polyimide membranes, *J. Membr. Sci.* 170 (2000) 205–214.

[114] J. Fang, H. Kita, K.I. Okamoto, Gas permeation properties of hyperbranched polyimide membranes, *J. Membr. Sci.* 182 (2001) 245–256.

[115] Y. Liu, R. Wang, T.S. Chung, Chemical cross-linking modification of polyimide membranes for gas separation, *J. Membr. Sci.* 189 (2001) 231–239.

[116] S. Matsui, H. Sato, T. Nakagawa, Effects of low molecular weight photosensitizer and UV irradiation on gas permeability and selectivity of polyimide membrane, *J. Membr. Sci.* 141 (1998) 31–43.

[117] K. Tanaka, A. Taguchi, J. Hao, H. Kita, K. Okamoto, Permeation and separation properties of polyimide membranes to olefins and paraffins, *J. Membr. Sci.* 121 (1996) 197–207.

[118] E.P. Favvas, G.E. Romanos, S.K. Papageorgiou, F.K. Katsaros, A.Ch. Mitropoulos, N.K. Kanellopoulos, A methodology for the morphological and physicochemical characterisation of asymmetric carbon hollow fiber membranes, *J. Membr. Sci.* 375 (2011) 113–123.

[119] P. Pandey, R.S. Chauhan, Membranes for gas separation, *Prog. Polym. Sci.* 26 (2001) 853–893.

[120] H.I. Mahon, Permeability separatory apparatus, permeability separatory membrane element, method of making the same and process utilizing the same, US Patent 3,228,876 (January 1966).

[121] C.W. Jones. W.J. Koros, Carbon molecular sieve gas separation membranes - I. preparation and characterization based on polyimide precursors, *Carbon* 32 (1994) 1419–1425.

[122] A.B. Fuertes, D.M. Nevskaia, T.A. Centeno, Carbon composite membranes from Matrimid® and Kapton® polyimides for gas separation, *Microp. Mesopor. Mater.* 33 (1999) 115–125.

[123] A.S. Ghosal, W.J. Koros, Air separation properties of flat sheet homogeneous pyrolytic carbon membranes, *J. Membr. Sci.* 174 (2000) 177–188.

[124] D.Q. Vu, W.J. Koros, S.J. Miller, Effect of condensable impurities in CO_2/CH_4 gas feeds on carbon molecular sieve hollow-fiber membranes, *Ind. Eng. Chem. Res.* 42 (2003) 1064–1075.

[125] D.Q. Vu, W.J. Koros, High pressure CO_2/CH_4 separation using carbon molecular sieve hollow fiber membranes, *Ind. Eng. Chem. Res.* 41 (2002) 367–380.

[126] E.P. Favvas, E.P. Kouvelos, G.E. Romanos, G.I. Pilatos, A Ch. Mitropoulos, N.K. Kanellopoulos, Characterization of highly selective microporous carbon hollow fiber membranes prepared from a commercial co-polyimide precursor, *J. Porous Mater.*, 15 (2008) 625–633.

[127] X. Ma, R. Swaidan, B. Teng, H. Tan, O. Salinas, E. Litwiller, Y. Han, I. Pinnau, Carbon molecular sieve gas separation membranes based on an intrinsically microporous polyimide precursor, *Carbon*, 62 (2013) 88–96.

[128] R. Swaidan, X. Ma, E. Litwiller, I. Pinnau, High pressure pure- and mixed-gas separation of CO_2/CH_4 by thermally-rearranged and carbon molecular sieve membranes derived from a polyimide of intrinsic microporosity, *J. Membr. Sci.* 447 (2013) 387–394.

[129] M.G. Sedigh, L. Xu, T.T. Tsotsis, M. Sahimi, Transport and morphological characteristics of polyetherimide-based carbon molecular sieve membranes, *Ind. Eng. Chem. Res.* 38 (1999) 3367–3380.

[130] W.N.W. Salleh, A.F. Ismail, Fabrication and characterization of PEI/PVP-based carbon hollow fiber membranes for CO_2/CH_4 and CO_2/N_2 separation, *AIChE* 58 (2012) 3167–3175.

[131] W.N.W. Salleh, A.F. Ismail, Carbon hollow fiber membranes derived from PEI/ PVP for gas separation, *Separ. Purif. Technol.* 80 (2011) 541–548.

[132] H.-C. Lee, M. Monji, D. Parsley, M. Sahimi, P. Liu, F. Egolfopoulos, T. Tsotsis, Use of steam activation as a post-treatment technique in the preparation of carbon molecular sieve membranes, *Ind. Eng. Chem. Res.* 52 (2013) 1122–1132.

[133] T.A. Centeno, A.B. Fuertes, Supported carbon molecular sieve membranes based on a phenolic resin, *J. Membr. Sci.* 160 (1999) 201–211.

[134] M.G. Sedigh, W.J. Onstot, L. Xu, W.L. Peng, T.T. Tsotsis, M. Sahimi, Experiments and simulation of transport and separation of gas mixtures in carbon molecular sieve membranes, *J. Phys. Chem. A* 102 (1998) 8580–8589.

[135] L.Y. Jiang, T.-S. Chung, R. Rajagopalan, Dual-layer hollow carbon fiber membranes for gas separation consisting of carbon and mixed matrix layers, *Carbon* 45 (2007) 166–172.

[136] S.S. Hosseini, T.-S. Chung, Carbon membranes from blends of PBI and polyimides for N_2/CH_4 and CO_2/CH_4 separation and hydrogen purification, *J. Membr. Sci.* 328 (2009) 174–185.

[137] X. He, J.A. Lie, E. Sheridan, M.-B.t Hägg, Preparation and characterization of hollow fiber carbon membranes from cellulose acetate precursors, *Ind. Eng. Chem. Res.* 50 (2011) 2080–2087.

[138] X. He, Y. Chu, A. Lindbråthen, M. Hillestad, M.-B. Hägg, Carbon molecular sieve membranes for biogas upgrading: Techno-economic feasibility analysis, *J. Clean. Prod.* 194 (2018) 584–593.

[139] L. Lei, A. Lindbråthen, M. Hillestad, M. Sandru, E. Favvas, X. He, Screening cellulose spinning parameters for fabrication of novel carbon hollow fiber membranes for gas separation, *Ind. Eng. Chem. Res.* 58 (2019) 13330–13339.

[140] A.D. Wiheeb, I.K. Shamsudin, M.A. Ahmad, M.N. Murat, J. Kim, M.R. Othman, Present technologies for hydrogen sulfide removal from gaseous mixtures, *Rev. Chem. Eng.* 29 (2013) 449–470.

[141] A.B. Jensen and C. Webb, Treatment of H_2S-containing gases: A review of microbiological alternatives, *Enzyme Microb. Technol.* 17 (1995) 2–10.

[142] K. Li, D. Wang, C.C. Koe, W.K. Teo, Use of asymmetric hollow fibre modules for elimination of H_2S from gas streams via a membrane absorption method, *Chem. Eng. Sci.* 53 (1998) 1111–1119.

[143] R. Faiz, M. Al-Marzouqi, Mathematical modeling for the simultaneous absorption of CO_2 and H_2S using MEA in hollow fiber membrane contactors, *J. Membr. Sci.* 342 (2009) 269–278.

[144] T. Kameyama, M. Dokiya, M. Fujishige, H. Yokokawa, K. Fukuda, Production of hydrogen from hydrogen sulfide by means of selective diffusion membranes, *Int. J. Hydrogen Energy* 8 (1983) 5–13.

[145] H. Ohashi, H. Ohya, M. Aihara, Y. Negishi, S.I. Semenova, Hydrogen production from hydrogen sulfide using membrane reactor integrated with porous membrane having thermal and corrosion resistance, *J. Membr. Sci.* 146 (1998) 39–52.

[146] K. Akamatsu, M. Nakane, T. Sugawara, T. Hattori, S.I. Nakao, Development of a membrane reactor for decomposing hydrogen sulfide into hydrogen using a high-performance amorphous silica membrane, *J. Membr. Sci.* 325 (2008) 16–19.

[147] R. Govind, D. Atnoor, Development of a composite palladium membrane for selective hydrogen separation at high temperature, *Ind. Eng. Chem. Res.* 30 (1991) 591–594.

[148] D.J. Edlund, W.A. Pledger, Catalytic platinum-based membrane reactor for removal of H2S from natural gas streams, *J. Membr. Sci.* 94 (1994) 111–119.

[149] H Neomagus, W. van Swaaij, G. Versteeg, The catalytic oxidation of H_2S in a stainless steel membrane reactor with separate feed of reactants, *J. Membr. Sci.* 148 (1998) 147–160.

[150] S. Weller, W.A. Steiner, Engineering aspects of separation of gases–fractional permeation through membranes. *Chem. Engin. Progress* 46 (1950) 585–590.

[151] S. Weller, W.A. Steiner, Separation of gases by fractional permeation through membranes. *J. Appl. Phys.* 21 (1950) 279–283.

[152] D.W. Brubaker, K. Kammermeyer, Separation of gases by plastic membranes–permeation rates and extent of separation, *Ind. Engin. Chem. Res.* 46 (1954) 733–739.

[153] Y. Shindo, T. Hakuta, H. Yoshitome, H. Inoue, Calculation methods for multicomponent gas separation by permeation, *Separ. Sci. Technol.* 20 (1985) 445–459.

[154] C.Y. Pan, Gas separation by high-flux, asymmetric hollow-fiber membrane, *AIChE J.* 32 (1986) 2020–2027.

[155] D.T. Coker, B.D. Freeman, G.K. Fleming, Modeling multicomponent gas separation using hollow-fiber membrane contactors, *AIChE J.* 44 (1998) 1289–1302.

[156] M. Scholz, T. Harlacher, T. Melin, M. Wessling, Modeling gas permeation by linking nonideal effects, *Ind. Eng. Chem. Res.* 52 (2012) 1079–1088.

[157] S.P. Kaldis, G.C. Kapantaidakis, T.I. Papadopoulos, G.P. Sakellaropoulos, Simulation of binary gas separation in hollow fiber asymmetric membranes by orthogonal collocation, *J. Membr. Sci.* 142 (1998) 43–59.

[158] R. Khalilpour, A. Abbas, Z.P. Lai, I. Pinnau, Modeling and parametric analysis of hollow fiber membrane system for carbon capture from multicomponent flue gas, *AIChE J.* 58 (2012) 1550–1561.

[159] R.W. Baker, K. Lokhandwala, Natural gas processing with membranes: An overview, *Ind. Engin. Chem. Res.* 47 (2008) 2109–2121.

[160] A. Alshehri, R. Khalilpour, A. Abasa, Z. Laib, Membrane systems engineering for post-combustion carbon capture, *Energy Procedia* 37 (2013) 976–985.

[161] T. Pettersen, K.M. Lien, Design studies of membrane permeator processes for gas separation, *Gas Separ. Purif.* 9 (1995) 151–169.

[162] M.S. Avgidou, S.P. Kaldis, G.P. Sakellaropoulos, Membrane cascade schemes for the separation of LPG olefins and paraffins, *J. Membr. Sci.* 233 (2004) 21–37.

[163] R. Carapellucci, A. Milazzo, Membrane systems for CO_2 capture and their integration with gas turbine plants, *Proceedings of the Institution of Mechanical Engineers A J. Power Energy* 217 (2003) 505–517.

[164] M.T. Ho, G.W. Allinson, D.E. Wiley, Reducing the cost of CO_2 capture from flue gases using membrane technology, *Ind. Eng. Chem. Res.* 47 (2008) 1562–1568.

[165] T.C. Merkel, H. Lin, X. Wei, R. Baker, Power plant post-combustion carbon dioxide capture: an opportunity for membranes, *J. Membr. Sci.* 359 (2010) 126–139.

[166] M. Stewart and K. Arnold, *Gas Sweetening and Processing Field Manual*, Gulf Professional Publishing, Amsterdam, 2011.

[167] M. Rezakazemi, I. Heydari, Z. Zhang, Hybrid systems: combining membrane and absorption technologies leads to more efficient acid gases (CO_2 and H_2S) removal from natural gas, *J. CO2 Util.* 18 (2017) 362–369.

[168] J.P. Vandersluijs, C.A. Hendriks, K. Blok, Feasibility of polymer membranes for carbon-dioxide recovery from flue-gases, *Energy Convers. Manage.* 33 (1992) 429–436.

[169] S. Kazama, S. Morimoto, S. Tanaka, H. Mano, T. Yashima, K. Yamada, K. Haraya, Cardo polyimide membranes for CO_2 capture from flue gases, *7th Intern. Conf. on Greenhouse Gas Control Technologies, Greenhouse Gas Control Technologies*, Vancouver, Vol. 1, 2004, 75–82.

[170] R. Bounaceur, N. Lape, D. Roizard, C. Vallieres, E. Favre, Membrane processes for post-combustion carbon dioxide capture: a parametric study, *Energy* 31 (2006) 2556–2570.

[171] M.T. Ho, G. Allinson, D.E. Wiley, Comparison of CO_2 separation options for geo-sequestration: are membranes competitive? *Desalination* 192 (2006) 288–295.

[172] E. Favre, Carbon dioxide recovery from post-combustion processes: can gas permeation membranes compete with absorption? *J. Membr. Sci.* 294 (2007) 50–59.

[173] L. Zhao, E. Riensche, R. Menzer, L. Blum, D. Stolten, A parametric study of CO_2/N_2 gas separation membrane processes for post-combustion capture, *J. Membr. Sci.* 325 (2008) 284–294.

[174] A. Hussain and M.-B. Hägg, A feasibility study of CO_2 capture from flue gas by a facilitated transport membrane, *J. Membr. Sci.* 359 (2010) 140–148.

[175] H. Zhai and E.S. Rubin, Techno-economic assessment of polymer membrane systems for postcombustion carbon capture at coal-fired power plants, *Environ. Sci. Technol.* 47 (2013) 3006–3014.

[176] F. Ahmad, K.K. Lau, A.M. Shariff, G. Murshid, Process simulation and optimal design of membrane separation system for CO_2 capture from natural gas, *Comput. Chem. Eng.* 36 (2012) 119–128.

[177] F. Ahmad, K.K. Lau, A.M. Shariff, Y. Fong Yeong, Temperature and pressure dependence of membrane permeance and its effect on process economics of hollow fiber gas separation system, *J. Membr. Sci.* 430 (2013) 44–55.

[178] L. Peters, A. Hussain, M. Follmann, T. Melin, M.-B. Hägg, CO_2 removal from natural gas by employing amine absorption and membrane technology—A technical and economical analysis, *Chem. Eng. J.* 172 (2011) 952–960.

[179] S.S.M. Lock, K.K. Lau, A.M. Shariff, Effect of recycle ratio on the cost of natural gas processing in countercurrent hollow fiber membrane system, *Ind. Eng. Chem. Res.* 21 (2015) 542–551.

[180] X. He, M.-B. Hägg, T.-J. Kim, Hybrid FSC membrane for CO_2 removal from natural gas: Experimental, process simulation, and economic feasibility analysis, *AIChE J.* 60 (2014) 4174–4184.

[181] S. Haider, A. Lindbrathen, M.-B. Hagg, Techno-economical evaluation of membrane based biogas upgrading system: A comparison between polymeric membrane and carbon membrane technology, *Green Energy Environ.* 1 (2016) 222–234.

[182] S. Haider, A. Lindbrathen, J. Lie, I. Andersen, M.-B. Hagg, CO_2 separation with carbon membranes in high pressure and elevated temperature applications, *Separ. Purif. Technol.* 190 (2018) 177–189.

[183] Y. Chu, X. He, Process simulation and cost evaluation of carbon membranes for CO_2 removal from high-pressure natural gas, *Membranes*, 8 (2018) 118.

7

Carbon Membranes for H_2 Purification

Miki Yoshimune

National Institute of Advanced Industrial Science and Technology (AIST)

7.1 Introduction

This chapter focuses on carbon membranes for use in hydrogen (H_2)-purification processes. These processes include the separation of gases, such as H_2/nitrogen (N_2), H_2/methane (CH_4), H_2/carbon dioxide (CO_2), and H_2/hydrocarbons. Compared with conventional polymeric membranes and other inorganic membranes, the advantages of carbon membranes include their high H_2 selectivities in addition to a high heat resistance and a high chemical resistance that is derived from the carbon material. The high H_2 selectivities of carbon membranes originate from the selective permeation of small H_2 molecules through the micropores (0.3–0.5 nm), and the separation performance of carbon membranes exceeds that of conventional polymeric membranes because of this molecular sieving effect.

Selective-surface-flow membranes that are derived from a vinylidene-chloride copolymer with larger micropores (0.5–0.7 nm) were developed by Rao and Sircar [1, 2]. Selective-surface-flow membranes show high selectivities toward more strongly adsorbed components in a gas mixture by selective adsorption and surface diffusion. Recently, graphene and graphene-oxide membranes for gas separation have been the focus of much attention, and some show high H_2 selectivities [3–5]. However, the preparation of these membranes is a multistep process and extensive effort is required to obtain defect-free membranes. This chapter focuses on carbon molecular sieve (CMS) membranes that have been developed over the past decade for possible practical use for H_2 separation.

7.2 Precursor Polymer Selection

Typically, CMS membranes are prepared by polymeric precursor pyrolysis. This process has been studied extensively since the 1980s. Pioneering works are summarized in excellent reviews and books [6–12]. Polymers such as cellulose derivatives, polyimides, polyetherimide (PEI), phenolic resins, polyfurfuryl alcohol (PFA), and polyphenylene oxide (PPO) have been used as precursors to produce CMS membranes. Polybenzimidazole (PBI), which is a polymer of intrinsic microporosity (PIM), poly(phthalazinone ether sulfone ketone) (PPESK), and polyester were developed as a new type of polymer precursor. Tables 7.1 and 7.2 summarize the H_2 permeation properties of reported carbon membranes that have been developed over the past decade according to their precursors and configurations. They are introduced in detail in the following section.

7.2.1 Cellulose Derivatives

He et al. [13] reported hollow-fiber carbon membranes that were prepared from different deacetylated cellulose-acetate precursors. The precursors were carbonized using CO_2 as the purge gas, a heating rate of 4°C/min, a final temperature of 650°C, and a final soak time of 2 h. Single-gas permeation tests with H_2, CO_2, O_2, N_2, and CH_4 were conducted, and the influences of temperature and feed pressure on the gas-separation performance of the membranes were studied. A gas-mixture test was also performed, which indicated that the permeability and selectivity should be optimized for the specific carbon-membrane-separation process.

Campo et al. [14] prepared CMS membranes from commercial cellophane films in a single heating step. Cellophane is a thin sheet of regenerated cellulose, which has advantages such as a low cost, ready availability, and biodegradability. The effects of pyrolysis temperature, soaking time, and pyrolysis atmosphere on the membrane performance were investigated extensively. Maximum permeability was reached for a membrane that was pyrolyzed at 550°C and its separation performance exceeded the Robeson's upper bound [34, 35]. Sá et al. [24] examined CMS membranes that were supplied by Carbon Membranes Ltd., Israel, as a membrane reactor (MR) for methanol steam reforming at 150 and 200°C. These membranes were produced by pyrolysis of cellulose-derived hollow fibers in an inert atmosphere, followed by several carbon chemical-vapor-deposition (CVD)/activation steps. The last step was performed at elevated temperature using air. This post treatment assigns the membranes a precise pore-size distribution that allows for very high permeabilities and selectivities. The single- and mixed-gas permeance measurements at 200°C were similar and a good agreement resulted

TABLE 7.1

Hydrogen-permeation properties (permeability) of carbon membranes developed over the past decade.

Precursor	Configuration	H₂ permeability (barrer)	Ideal selectivity			Reference
			H_2/CO_2	H_2/N_2	H_2/CH_4	
Cellulose acetate	Hollow fiber	~1000	~3	~20	~40	He et al. [13]
Cellophane	Flat	168.1	9.9	391	989	Campo et al. [14]
PEI+SBA-15	Disk	667.5	3.0	75.0	75.0	Tseng et al. [15]
PEI	Disk	600.7	8.3	314	725.9	Tseng et al. [16]
Phenolic resin	Tube	2047	1.8	65.6		Teixeira et al. [17]
Resorcinol–formaldehyde resin	Tube	445.6		>586		Rodrigues et al. [18]
PPO/PVP	Disk	1121		163.9	160.9	Itta et al. [19]
PPO/PI	Disk	1448.3	1.1	171.9	172.4	Tseng and Itta [20]
PPO/PEI	Disk	811.6	1.6	136.2	136.3	
PBI/Torlon	Flat	970.3	3.47	155.5	343	Hosseini and Chung [21]
PBI/Matrimid		660.2	6.84	174.6	1396	
		324.0	8.85	257.1	1165	
		182.7	16.61	466.1	937	
PIM-6FDA-OH	Flat	5248	1.04	28	40	Ma et al. [22]
		4693	1.63	38	81	
		2177	3.92	128	363	
SPPESK	Flat	200	1.7	80		Zhang et al. [23]
		240	1.5	38		

between the separation factor and the ideal selectivity. The effect of the MR on the methanol steam-reforming reaction is described in Section 7.3.

7.2.2 Polyimides

Polyimides are excellent precursors because of their high glass transition temperatures (T_g), ease of processing, and the good separation performance of their CMS membranes [12]. From a practical perspective, Matrimid (a polyimide that is prepared from 3,3′, 4,4′-benzophenone tetracarboxylic

TABLE 7.2

Hydrogen-permeation properties (permeance) of carbon membranes developed over the past decade.

Precursor	Configuration	H$_2$ permeance (GPU)	Ideal selectivity			Reference
			H$_2$/CO$_2$	H$_2$/N$_2$	H$_2$/CH$_4$	
Cellulose derivative	Hollow fiber	120.2	14.5			Sá et al. [24]
Matrimid	Tube	29.4	2.4	4.5	10.5	Briceño et al. (2012a)
BTDA-TDI/ MDI (P84)	Hollow fiber	5.99	14.3	599.0	704.7	Favvas et al. [25]
		4.41	20.9	603	5500	
BTDA-TDI/ MDI (P84)	Tube	1399.7		434.7		Sazali et al. [26]
PMDA–ODA	Disk	163	6.2	76.3		Li et al. [27]
Commercial polyimide	Tube	3289	24	130	228	Ngamou et al. [28]
PFA	Tube	32.4	9.8	124.7	216.9	Song et al. [29]
PFA (FFA)	Tube	47.8	4.1		3.4	Hirota et al. [30]
SPPO	Hollow fiber	167	3.3	141	392	Yoshimune and Haraya [31]
		60	4.1	170	558	
SPPO + CVD	Hollow fiber	95	11.1	296	1886	Yoshimune and Haraya [32]
		62	69	2213	29941	
Polyester	Tube	1854	8.4	84	356	Richter et al. [33]

dianhydride (BTDA) and 4',6-diamidino-2-phenylindole (DAPI)) and P84 (a polyimide that is prepared from BTDA and 80% methylphenylenediamine (TDI) + 20% methylenediamine (MDI)) are sometimes used as precursors of commercially available polyimides. These polyimides can be cast into any form because they are soluble in various solvents.

Briceño et al. [36] pyrolyzed Matrimid that was coated on tubular ceramic supports at 550, 650, and 700°C to fabricate CMS membranes for H$_2$ separation. After the preparation conditions had been optimized, the obtained membrane exhibited ideal selectivities of 2.37, 4.70, and 10.62 for H$_2$/CO$_2$, H$_2$/CO, and H$_2$/CH$_4$, respectively. Permeance values exceeded 9.82×10^{-9} mol/(m^2 Pa s) during pure H$_2$-permeation tests. Favvas et al. [25] produced hollow-fiber CMS membranes by P84 hollow-fiber pyrolysis under Ar at 900°C for He and H_2 separation. The effect of pyrolysis isothermal time was investigated and membranes that remained at the higher pyrolysis temperature for a longer time exhibited lower permeance values. As a result, these membranes exhibit excellent separation factors for H_2/CH_4 and H_2/N_2 of 5500 and 600, respectively. Furthermore, gas-mixture experiments were performed, which resulted in effective separation factors in He/CH_4 and H_2/CH_4 mixtures. Sazali et al. [26] produced carbon tubular membranes from P84 co-polyimide as a precursor material and nanocrystalline cellulose (NCC) as a pore-forming agent for H_2 and He separation. The results showed that higher carbonization temperatures resulted in more selective but less productive carbon membranes. The presence of NCC as an additive provides the best pore structural properties. The carbon membrane carbonized at 800°C with 7 wt% NCC additions showed the most promising H_2/N_2 selectivity of 434.68 ± 1.39. Li et al. [27] coated poly(amic acid) precursors derived from pyromellitic dianhydride and 4,4'-oxydianiline on carbon disks with or without the ordered mesoporous carbon interlayer to prepare CMS membranes. The ordered mesoporous carbon interlayer can reduce support surface defects, improve the interfacial adhesion of the support to the thin separation layer, and enhance the gas-permeation properties of the supported CMS membranes. The CMS membranes achieved a H_2 permeance of 545.5 × 10^{-10} mol m^{-2} s^{-1} Pa^{-1} and exhibited ideal selectivities of 76.3 for H_2/N_2.

Ngamou et al. [28] fabricated ultrathin (~200 nm) and pinhole-free CMS membranes by carbonization of a commercial polyimide resin at 700°C on the inner surface of hierarchically structured porous γ-Al$_2$O$_3$/α-Al$_2$O$_3$ supports. The synthesized CMS membranes showed an unprecedentedly high H_2 permeance of up to 1.1 × 10^{-6} mol m^{-2} s^{-1} Pa^{-1} and ideal separation factors of 24, 130, and 228 for H_2/CO_2, H_2/N_2, and H_2/CH_4, respectively, at 200°C. These are the highest reported separation performances for carbon membranes, and the developed CMS membranes show potential for high-temperature H_2 purification.

7.2.3 Polyetherimides

Polyetherimide is a commercially available precursor that has similar properties to polyimides but is less expensive than polyimide-based material. Tseng et al. [15] used commercial polyetherimide as a polymeric precursor to prepare CMS membranes on a porous α-alumina support disk for H_2 separation. To improve gas permeance without losing selectivity, mesoporous silica, SBA-15, was incorporated into the polymer precursor and its effect on the gas-transport properties of the resultant CMS membranes evaluated. The SBA-15/CMS composite membrane showed higher gas permeabilities than the pure CMS membrane and high selectivities were obtained only in the 0.5 wt% SBA-15 loading CMS membrane.

Tseng et al. [16] synthesized PEI-derived CMS membranes on titanium-gel-modified alumina supports and the effects of the TiO_2 nanonetwork physicochemical properties on the CMS layer texture and gas-separation performance of the CMS membrane were investigated. The TiO_2 intermediate layer controls the interlocking pattern between the selective layer and the porous Al_2O_3 support. The CMS membrane supported on the bare Al_2O_3 substrate had a H_2 permeability of 537.5 barrer and a H_2/CH_4 selectivity of 197.6. After modifying the CMS membrane that was supported on a TiO_2/Al_2O_3 substrate, a superior H_2/CH_4 selectivity of 725.9 was exhibited, with an increased H_2 permeability of 600.7 barrer.

7.2.4 Phenolic Resins and Analogous Compounds

Phenolic resins, which are very popular, inexpensive, and high-carbon-yield polymers, have also been used as precursors to prepare CMS membranes. Phenolic resins form self-standing membranes with difficulty, so the CMS membranes are prepared on supports.

Teixeira et al. [17] prepared composite CMS membranes from resol phenolic resin solutions (12.5–15 wt%) loaded with boehmite nanoparticles (0.5–1.2 wt%) on tubular Al_2O_3 supports. After pyrolysis at 550°C for 2 h, a carbon matrix with well-dispersed Al_2O_3 nanowires was formed from resin decomposition and boehmite dehydroxylation. CMS membranes with higher carbon/Al_2O_3 ratios had a more open porous structure and exhibited higher gas permeabilities and lower selectivities; for H_2, the permeability was 2047 barrer and the H_2/N_2 selectivity was 65.6.

Resorcinol–formaldehyde resin serves as an excellent precursor material for CMS membrane production because of its considerable carbon yield, high inherent purity, and low cost, like phenolic resins. Rodrigues et al. [18] prepared CMS membranes on α-Al_2O_3 tubular supports by carbonization of a resorcinol–formaldehyde resin loaded with boehmite. CMS membranes that were carbonized at 550°C showed a large number of micropores with a more narrow pore-size distribution, much higher ideal selectivities, and relatively similar gas-permeation rates compared with those produced at 500°C, for which the H_2 permeability is 301 barrer and the H_2/N_2 selectivity exceeds 586.

7.2.5 Polyfurfuryl Alcohol

PFA has been used as a precursor for CMS membranes for many years. Because it is liquid at room temperature, all membranes derived from PFA are composite membranes that are supported by porous substrates.

Song et al. [29] produced composite CMS membranes by the direct coating of a layer of PFA that was polymerized by an iodine catalyst onto a porous coal-based carbon tubular support followed by controlled pyrolysis at 600–900°C. The uniform and defect-free thin top layer can be prepared using PFA liquid in one coating step. The highest H_2 permeance, 108.45×10^{-10} mol m^{-2} s^{-1} Pa^{-1}, was obtained at 600°C and the H_2/N_2 ideal selectivity was 124.72. Hirota et al. [30] prepared H_2-permselective CMS membranes by a vapor-phase synthesis using furfuryl alcohol (FFA) as a carbon source. The pore size of the FFA CMS membranes was increased by post-synthesis activation using various gases and vapors such as H_2, CO_2, O_2, and steam. After activation in H_2 and steam, the pore size of the CMS membrane increased from 0.3 to 0.45 nm, which resulted in an increase in H_2 permeance from 3.6×10^{-9} to 1.6×10^{-8} mol m^{-2} s^{-1} Pa^{-1} with high permselectivities of H_2/CF_4 (>1500). The activated FFA CMS membrane was applied to the dehydrogenation of methylcyclohexane in a MR, as described in Section 7.3.

7.2.6 Polyphenylene Oxide

Poly (2,6-dimethyl-1,4-phenylene oxide) (poly(phenylene oxide); PPO) is a versatile commercial polymer that is used as an engineering plastic; it is much cheaper than polyimides. PPO is thermally stable, with a higher glass transition temperature, and it is easy to modify with functional groups to achieve more favorable gas separation. Since Yoshimune et al. [37] produced novel CMS membranes from PPO and its derivatives, PPO has been chosen as a polymer precursor to produce CMS membranes.

Itta et al. [19] fabricated CMS membranes from PPO with the thermally labile polymer polyvinylpyrrolidone (PVP) coated on alumina disks using a spin-coating technique, for H_2/N_2 and H_2/CH_4 separation. The effect of PPO content on the membrane performance was evaluated. The best performance for H_2 permeability, 1121 barrer, was obtained from a solution of 12.5 wt% PPO with 2.5 wt% PVP that was pyrolyzed at 700°C: the H_2/N_2 and H_2/CH_4 selectivities were 163.9 and 160.9, respectively. The same group also investigated the pore-size control of PPO-derived CMS membranes by PI and PEI addition to precursor solutions [20]. After modification, the H_2-separation performance of these CMS membranes was higher than the Robeson's upper bound curves. Compared with unmodified PI-derived CMS membranes, the H_2 permeability of the PPO/PI-derived CMS membranes increased from 565 to 1448 barrer and the ideal H_2/N_2 selectivity increased from 17.1 to 171.8.

Yoshimune and Haraya [31] developed flexible hollow-fiber CMS membranes by using sulfonated PPO (SPPO) as a precursor polymer from a practical perspective to produce large-scale membrane modules. A lower pyrolysis temperature (450–600°C) gives more flexible CMS membranes with smaller bending radii. The SPPO hollow-fiber CMS membrane that was pyrolyzed at 600°C showed the best gas-separation performance with sufficient membrane flexibility, with a H_2 permeance and H_2/CH_4 selectivity of 167 GPU and 392, respectively. Recently, Yoshimune and Haraya [32] investigated the effects of one-step CVD of propylene on the pore structures and gas-separation performances of hollow-fiber CMS membranes that were derived from

SPPO. The CVD carbon membranes showed lower permeances but higher selectivities compared with those prepared without CVD, which indicates that the effective pore size of the CVD carbon membranes decreased with CVD. The gas-separation performance could be controlled easily by changing the propylene volume fraction and CVD time. The highest H_2/CH_4 selectivity, 29 940, was obtained for a membrane that was pyrolyzed at 700°C for 20 min with CVD by using 10% propylene for 20 min.

7.2.7 Other Polymer Precursors

Hosseini and Chung [21] studied flat CMS membranes that were derived from PBI with various polyimides, such as Matrimid 5218, Torlon 4000T, and P84, to explore the advantages of using blending techniques to enhance the gas-separation performance of CMS membranes. The CMS membranes that were derived from PBI/Matrimid exhibited a more attractive performance than the other blends when pyrolyzed at 800°C. Chemical modification using *p*-xylene diamine as the chemical agent for crosslinking PBI/Matrimid precursors prior to carbonization enhanced the H_2/N_2 and H_2/CO_2 separation performance.

Ma et al. [22] developed a new class of precursors of an intrinsically microporous polyimide that contained spirocenters (PIM-6FDA-OH). PIMs are easily solution-processable and durable membrane materials with a high Brunauer–Emmett–Teller (BET) surface area (up to 1000 m^2/g). Gradual changes in the membrane microstructure and gas-transport properties were studied by stepwise heat treatment at 440, 530, 600, 630, and 800°C. At 440°C, the PIM-6FDA-OH was converted to a polybenzoxazole and exhibited a twofold increase in H_2 permeability (from 259 to 578 barrer). At 530°C, a distinct intermediate amorphous carbon structure with superior gas-separation properties was formed and an 11-fold increase in H_2 permeability resulted (2860 barrer) over the pristine polymer. The graphitic carbon membrane, which was obtained by heat treatment at 600°C, exhibited excellent gas-separation properties, including a remarkable H_2 permeability of 5248 barrer with a high selectivity over CH_4 of 40. When pyrolyzed above 600°C, a decrease in permeability was observed because of pore shrinkage, which yielded a significant molecular sieving effect, and the CMS membranes that were obtained at 800°C exhibited an excellent gas-separation performance. Zhang et al. [23] prepared CMS membranes that were derived from sulfonated PPESK as a novel polymeric precursor and investigated the effects of sulfonation degree (SD) on the gas-permeation properties. The CMS membrane microstructure had a higher pore volume with an increase in the SD from 59 to 75% because of pendent sulfonic-acid-group decomposition during heat treatment. The H_2 permeability of the CMS membrane increased from 200 to 240 barrer, with a decrease of 80 to 38 in H_2/N_2 selectivity. The SD in the SPPSK increased from 59 to 75%.

Richter et al. [33] obtained a 125-nm-thick carbon layer on the inner surface of a porous γ-Al$_2$O$_3$/α-Al$_2$O$_3$ tube by controlled polyester carbonization at 500°C. The precursor solution was prepared by reaction of a propanediol with maleic-acid anhydride and phthalic-acid anhydride to give an unsaturated linear polyester and subsequent addition of styrene as cross-linker. For single-gas permeation, the CMS membrane showed a high H$_2$/C$_3$H$_8$ selectivity of 10000 with a H$_2$ permeance of 5 m^3 (STP)/(m^2 h bar). With a mixed feed of 80 vol% H$_2$/C$_3$H$_8$ at 2.55 bar, the permeate at 1.05 bar consisted of 99.99% H$_2$ with a high H$_2$ permeance of 4.5 m^3 (STP)/(m^2 h bar). This separation performance remained stable for more than one month under industrial conditions at BASF, for a 50 vol% H$_2$/C$_3$H$_8$ mixture at 300°C and 2.64 bar feed pressure with a stable H$_2$ permeance of 9.6 m^3 (STP)/(m^2 h bar).

7.3 Applications of Carbon Membranes in Hydrogen Separation

Most previous work has focused on developing high-performance CMS membranes on the basis of laboratory-scale gas-permeation measurement for H$_2$ separation, and few commercial CMS-membrane H$_2$-separation processes have been reported. CMS membranes provide great potential for selected gas separation, especially of H$_2$, because of their mechanical and chemical stabilities with superior separation performances. A potential application for CMS membranes in the recovery of H$_2$ that is transmitted with natural gas was evaluated by Grainger and Hägg [38]. Performance data for cellulose-derived CMS membranes were applied to a techno-economic evaluation of H$_2$ recovery and the results were compared with a commercial polyimide membrane as a benchmark. The CMS membranes produced higher-purity H$_2$, consumed less energy during separation, and achieved competitive specific separation costs under certain conditions. The CMS membranes easily achieved single-stage separation specifications at higher permeate pressures compared with the polyimide membrane because the H$_2$ selectivity of CMS membranes was higher. He [39] conducted a techno-economic feasibility analysis on carbon membranes for H$_2$ purification from a biomass fermentation process. A novel, energy-efficient, two-stage carbon membrane system for H$_2$ purification based on a combination of H$_2$-selective CMS membranes in the first stage and CO$_2$-selective carbon membranes in the second stage was proposed. The designed two-stage carbon-membrane system can capture CO$_2$ >95 vol% in the first-stage membrane unit at 20 bar and 120°C, and the high-purity H$_2$ (>99.5 vol%) with a high H$_2$ recovery (>95%) was produced in the second-stage membrane unit that was operated at a feed operation pressure of 20 bar and 20°C.

CMS membranes can be considered as possible alternatives to Pd-based membranes for use in MR H_2 production. Pd-based membranes are expensive and experience H_2-embrittlement cracking during thermal cycling and surface contamination by sulfur-containing species. CMS membranes have clear advantages in terms of their chemical stability, lower production costs, and relatively high selectivity for H_2. However, they are fragile and cannot be used in oxidizing atmospheres, so only a few applications of carbon membranes in MRs have been reported. Table 7.3 summarizes examples of the use of CMS MRs for H_2-related reactions.

The first experimental CMS MR was reported by Itoh and Haraya [40]. The MR for cyclohexane dehydrogenation consisted of a bundle of 20 hollow carbon fibers that were produced by pyrolysis of hollow polyimide fibers that contained 0.5 wt% Pt/Al_2O_3 pellets as a catalyst. Figure 7.1 provides a schematic that shows the CMS MR that was developed in this study. Cyclohexane dehydrogenation to benzene was carried out at 195°C under atmospheric pressure. The CMS MR produced a conversion that was somewhat better than the equilibrium conversion; the findings were supported by a mathematical model for a limited range of reaction conditions.

Sznejer and Sheintuch [41] tested a carbon MR (CMR) for isobutane dehydrogenation at high temperatures (450–500°C) on chromia/alumina catalyst

TABLE 7.3

Examples of hydrogen-related carbon molecular sieve membrane reactors.

Precursor	Catalyst	Reaction	Temperature	Reference
Polyimide	Pt/Al_2O_3	Cyclohexane dehydrogenation	195°C	Itoh and Haraya [40]
Cellulose	Cr_2O_3/Al_2O_3	Isobutane dehydrogenation	~500°C	Sznejer and Sheintuch [41]
Phenolic resin + PEG	$Cu/ZnO/Al_2O_3$	Methanol steam reforming	~250°C	Zhang et al. [42]
Not disclosed	$CuO/ZnO/Al_2O_3$	Water-gas-shift reaction	250°C	Harale et al. [43]
PEI	Sulfided $Co/Mo/Al_2O_3$	Water-gas-shift reaction	~300°C	Abdollahi et al. [44]
Cellulose derivative	$CuO/ZnO/Al_2O_3$	Methanol steam reforming	~200°C	Sá et al. [24]
Matrimid	$CuO/ZnO/Al_2O_3$	Methanol steam reforming	~250°C	Briceňo et al. (2012b)
PFA (FFA)	Pt/Al_2O_3	Methylcyclohexane dehydrogenation	~220°C	Hirota et al. [30]
Phenolic resin-based polymer	CoS_2/MoS_2	Water-gas-shift reaction	~250°C	Parsley et al. [45]

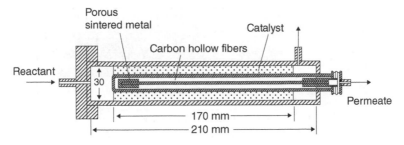

FIGURE 7.1
Schematic of a carbon membrane reactor [40].

pellets. The membrane module, which consisted of 100 hollow carbon fibers (Carbon Membranes Ltd.), had a H_2-to-isobutene permeability ratio above 100. Although the results obtained were better than those achieved with a corresponding fixed-bed reactor, the authors concluded that the improvement resulted because of sweeping N_2 transport and dilution. Simulations of the MR behavior showed poor agreement with the experiments [46]. Zhang et al. [42] studied the use of carbon membranes in methanol steam-reforming reactors. The carbon membrane was prepared from a novolac-type phenolic resin and poly(ethylene glycol) on a green support. Methanol steam reforming was performed at 200–250°C with a $Cu/ZnO/Al_2O_3$ catalyst, and the CMR was compared with a conventional fixed-bed reactor. A higher methanol conversion and lower carbon monoxide yield were achieved by enhancing the potential of the carbon membrane.

Harale et al. [43] used a CMR in the water-gas-shift reaction for H_2 production. The authors proposed a hybrid adsorbent MR (HAMR) system that combines the reaction and membrane-separation steps with adsorption. $CuO/ZnO/Al_2O_3$ was used as a catalyst, and a layered double hydroxide was selected as the CO_2 adsorbent. The carbon membranes, which were 25.4 cm long and had an outside diameter of 0.57 cm, showed high H_2 permeation fluxes at 250°C, but their preparation methods were not discussed. The experimental results agreed well with the model predictions, and the HAMR system can provide improved yields of H_2 with reduced CO concentrations. Abdollahi et al. [44] proposed the "one-box" process, which combines reaction and membrane separation in the same unit for the water-gas-shift reaction. It includes a catalytic MR, which makes use of a H_2-selective CMS membrane and a sulfur-tolerant $Co/Mo/Al_2O_3$ catalyst. The MR performance was investigated for a range of pressures and sweep ratios and showed higher CO conversions and a higher H_2 purity compared with the traditional packed-bed reactor. This result was extended to a field test by using a multitubular-supported CMS module with 86 single tubes and an effective membrane area of 0.76 m², shown in Figure 7.2, fabricated by Parsley et al. [45]. This module was tested above 250°C, and the membrane

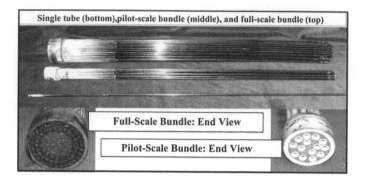

FIGURE 7.2
Photographs of a single tube and pilot-scale and full-scale bundles of carbon molecular sieve membranes [45].

performance remained unchanged over several hundred cumulative hours in the field test.

Sá et al. [24] studied the methanol steam-reforming reaction in a CMR over a commercial $CuO/ZnO/Al_2O_3$ catalyst at 150 and 200°C. CMS membranes prepared from cellulose derivative were supplied by Carbon Membranes Ltd. The CMR was operated at atmospheric pressure and with vacuum at the permeate side. It was found that methanol conversion, H_2 recovery, and H_2 yield were enhanced by lower feed flow rates because of higher residence times, with the drawback of higher carbon monoxide production. The simulation study showed that using water as sweep gas brings several advantages. In addition to an increase in methanol conversion and H_2 recovery, carbon monoxide production decreases significantly. Briceño et al. [47] applied CMS membranes to a methanol steam-reforming reaction in a MR at 550°C. The CMS membranes that were derived from Matrimid were pyrolyzed at 550°C with a H_2/N_2 ideal separation factor of 2.67–2.77 in the range of 23–150°C. There was little difference between a conventional reactor and MR because of the low H_2 selectivity; however, the total yield and methanol conversion were higher in the MR.

Hirota et al. [30] applied an activated FFA CMS membrane to methylcyclohexane dehydrogenation in a MR. The reaction temperatures were 200 and 220°C and 0.5 wt% Pt/Al_2O_3 was used as a catalyst. The methylcyclohexane conversion exceeded the equilibrium values because of the selective H_2 permeation.

Sá et al. [48] evaluated the potential advantages of a CMR by using a one-dimensional mathematical model compared with a Pd-MR for H_2 production by methanol steam reforming. The study focused on an analysis of the methanol conversion, the selectivity of the H_2/CO reaction, the CO concentration at the permeate side, and the H_2 recovery, and concluded that the CMR gave a higher H_2 recovery than the Pd-MR at high H_2 concentrations, but the Pd-MR showed more advantages at lower H_2 production rates. For successful use in MRs, CMRs require a high separation selectivity and a high permeability so that the permeation rate is comparable with the catalytic reaction rate. The key challenges in this context are to reduce the membrane thickness without introducing defects and to scale up techniques for carbon-membrane fabrication with large surface areas. For commercial applications, carbon membranes should be prepared as monoliths or hollow-fiber modules to provide the additional benefits of a low pressure drop and a high surface-to-volume ratio. It would also be advantageous to shift the thermodynamic equilibrium by improving the porous structure of the carbon membranes.

7.4 Summary

Recent technological developments in various CMS membranes for H_2 separation have been discussed in this chapter. Although few studies have examined H_2-separation processes using carbon membranes, CMS membranes offer several advantages, especially for MRs, such as a high resistance to chemicals and the capacity to operate at high temperatures. For CMS membrane commercialization, scaling up is one of the most challenging hurdles to surmount. It will be necessary to prepare high-quality CMS membranes with large surface areas in a reliable and cost-effective manner as well as to integrate these membranes into process modules with high-temperature sealing. The scope of this challenge is considerable, but extensive potential and opportunities exist for using new polymer precursors, new production methods, or new module-fabrication techniques. Proper reactor design, with regards to heat and mass transport issues during the reaction as well as the separation processes, is a significant factor.

References

[1] Rao, M.B., Sircar, S., 1993. Nanoporous carbon membranes for separation of gas mixtures by selective surface flow. *Journal of Membrane Science*, 85, 253–264.

[2] Rao, M.B., Sircar, S., 1996. Performance and pore characterization of nanoporous carbon membrane for gas separation. *Journal of Membrane Science*, 110, 109–118.

[3] Huang, S., Dakhchoune, M., Luo, W., Oveisi, E., He, G., Rezaei, M., Zhao, J., Alexander, D.T.L., Züttel, A., Strano, M.S., Agrawa, K.V., 2018. Single-layer graphene membranes by crack-free transfer for gas mixture separation. *Nature Communications*, 9, 2632–2642.

[4] Li, H., Song, Z., Zhang, X., Huang, Y., Li, S., Mao, Y., Ploehn, H.J., Bao, Y., Yu, M., 2013. Ultrathin, molecular-sieving graphene oxide membranes for selective hydrogen separation. *Science* 342, 95–98.

[5] Yoo, B.M., Shin, J.E., Lee, H.D., Park, H.B., 2017. Graphene and graphene oxide membranes for gas separation applications. *Current Opinion in Chemical Engineering*, 16, 39–47.

[6] Hamm, J.B., Ambrosi, A., Griebeler, J.G., Marcilio, N.R., Tessaro, I.C., Pollo, L.D., 2017. Recent advances in the development of supported carbon membranes for gas separation. *International Journal of Hydrogen Energy*, 42, 24830–24845.

[7] Ismail, A.F., David, L.I.B., 2001. A review on the latest development of carbon membranes for gas separation. *Journal of Membrane Science*, 193, 1–18.

[8] Ismail, A.F., Li, K., 2008. From polymeric precursors to hollow fiber carbon and ceramic membranes. *Inorganic Membranes: Synthesis, Characterization and Applications*. Mallada, R., Menendez, M. (Eds.), 81–120, Elsevier, Inc., Oxford.

[9] Kita, H., 2006. Gas and vapor separation membranes based on carbon membranes. *Material Science of Membranes for Gas and Vapor Separation*. Yampolskii, Y., Pinnau, I., Freeman, B.D. (Eds.), 337–354, John Wiley & Sons, Inc., Chichester.

[10] Salleh, W.N.W., Ismail, A.F., Matsuura, T., Abdullah, M.S., 2011. Precursor selection and process conditions in the preparation of carbon membrane for gas separation: A review. *Separation and Purification Reviews*, 40, 261–311.

[11] Saufi, S.M., Ismail, A.F., 2004. Fabrication of carbon membranes for gas separation—A review. *Carbon*. 42, 241–259.

[12] Williams, P.J., Koros, W.J., 2008. Gas separation by carbon membranes. *Advanced Membrane Technology and Applications*. Li, N.N., Fane, A.G., Ho, W.S.W., Matsuura, T. (Eds.), 599–632, John Wiley & Sons, Inc., New Jersey.

[13] He, X., Lie, J.A., Sheridan, E., Hägg, M.-B., 2011. Preparation and characterization of hollow fiber carbon membranes from cellulose acetate precursors. *Industrial & Engineering Chemistry Research*, 50, 2080–2087.

[14] Campo, M.C., Magalhães, F.D., Mendes, A., 2010. Carbon molecular sieve membranes from cellophane paper. *Journal of Membrane Science*, 350, 180–188.

[15] Tseng, H.-H., Shih, K., Shiu, P.-T., Lin, Y.-S., 2011. Effect of mesoporous silica modification on the structure of hybrid carbon membrane for hydrogen separation. *International Journal of Hydrogen Energy*, 36, 15352–15363.

[16] Tseng, H.-H., Wang C.-T., Zhuang G.-L., Uchytil P., Reznickova J., Setnickova K., 2016. Enhanced H_2/CH_4 and H_2/CO_2 separation by carbon molecular sieve membrane coated on titania modified alumina support: Effects of TiO_2 intermediate layer preparation variables on interfacial adhesion. *Journal of Membrane Science*, 510, 391–404.

[17] Teixeira, M., Rodrigues, S.C., Campo, M., Pacheco Tanaka, D.A., Llosa Tanco, M.A., Madeira, L., Sousa, J.M., Mendes, A., 2014. Boehmite-phenolic resin carbon molecular sieve membranes—Permeation and adsorption studies. *Chemical Engineering Research and Design*, 92, 2668–2680.

[18] Rodrigues, S.C., Whitley, R., Mendes, A., 2014. Preparation and characterization of carbon molecular sieve membranes based on resorcinol–formaldehyde resin. *Journal of Membrane Science*, 459, 207–216.

[19] Itta, A.K., Tseng, H.-H., Wey, M.-Y., 2011. Fabrication and characterization of PPO/PVP blend carbon molecular sieve membranes for H_2/N_2 and H_2/CH_4 separation. *Journal of Membrane Science*, 372, 387–395.

[20] Tseng, H.-H., Itta, A.K., 2012. Modification of carbon molecular sieve membrane structure by self-assisted deposition carbon segment for gas separation. *Journal of Membrane Science*, 389, 223–233.

[21] Hosseini, S.S., Chung, T.S., 2009. Carbon membranes from blends of PBI and polyimides for N_2/CH_4 and CO_2/CH_4 separation and hydrogen purification. *Journal of Membrane Science*, 328, 174–185.

[22] Ma, X., Swaidan, R., Teng, B., Tan, H., Salinas, O., Litwiller, E., Han, Y., Pinnau, I., 2013. Carbon molecular sieve gas separation membranes based on an intrinsically microporous polyimide precursor. *Carbon* 62, 88–96.

[23] Zhang, B., Wu, Y., Wang, T., Qiu, J., Zhang, S., 2011. Microporous carbon membranes from sulfonated poly(phthalazinone ether sulfone ketone): Preparation, characterization, and gas permeation. *Journal of Applied Polymer Science*, 122, 1190–1197.

[24] Sá, S., Sousa, J.M., Mendes, A., 2011. Steam reforming of methanol over a CuO/ZnO/Al₂O₃ catalyst part II: A carbon membrane reactor. *Chemical Engineering Science*, 66, 5523–5530.

[25] Favvas, E.P., Heliopoulos, N.S., Papageorgiou, S.K., Mitropoulos, A.C., Kapantaidakis, G.C., Kanellopoulos, N.K., 2015. Helium and hydrogen selective carbon hollow fiber membranes: The effect of pyrolysis isothermal time. *Separation and Purification Technology*, 142, 176–181.

[26] Sazali, N., Salleh, W.N.W., Ismail, A.F., 2017. Carbon tubular membranes from nanocrystalline cellulose blended with P84 co-polyimide for H₂ and He separation. *International Journal of Hydrogen Energy*, 42, 9952–9957.

[27] Li, L., Song, C., Jiang, H., Qiu, J., Wang, T., 2014. Preparation and gas separation performance of supported carbon membranes with ordered mesoporous carbon interlayer. *Journal of Membrane Science*, 450, 469–477.

[28] Ngamou, P.H.T., Ivanova, M.E., Guillon, O., Meulenberg, W.A., 2019. High-performance carbon molecular sieve membranes for hydrogen purification and pervaporation dehydration of organic solvents. *Journal of Materials Chemistry A*, 7, 7082–7091.

[29] Song, C., Wang, T., Qiu, J., 2009. Preparation of C/CMS composite membranes derived from poly(furfuryl alcohol) polymerized by iodine catalyst. *Desalination* 249, 486–489.

[30] Hirota, Y., Ishikado, A., Uchida, Y., Egashira, Y., Nishiyama, N., 2013. Pore size control of microporous carbon membranes by post-synthesis activation and their use in a membrane reactor for dehydrogenation of methylcyclohexane. *Journal of Membrane Science*, 440, 134–139.

[31] Yoshimune, M., Haraya, K., 2010. Flexible carbon hollow fiber membranes derived from sulfonated poly(phenylene oxide). *Separation and Purification Technology*, 75, 193–197.

[32] Yoshimune, M., Haraya, K., 2019. Simple control of the pore structures and gas separation performances of carbon hollow fiber membranes by chemical vapor deposition of propylene. *Separation and Purification Technology*, 233, 162–167.

[33] Richter, H., Voss, H., Kaltenborn, N., Kamnitz, S., Wollbrink, A., Feldhoff, A., Caro, J., Roitsch, S., Voigt, I., 2017. High-flux carbon molecular sieve membranes for gas separation. *Angewandte Chemie International Edition*, 56, 7760–7763.

[34] Robeson, L.M., 1991. Correlation of separation factor versus permeability for polymeric membranes. *Journal of Membrane Science*, 62, 165–185.

[35] Robeson, L.M., 2008. The upper bound revisited. *Journal of Membrane Science*, 321, 390–400.

[36] Briceño, K., Montané, D, Garcia-Valls, R., Iulianelli, A., Basile, A., 2012a. Fabrication variables affecting the structure and properties of supported carbon molecular sieve membranes for hydrogen separation. *Journal of Membrane Science*, 415–416, 288–297.

[37] Yoshimune, M., Fujiwara, I., Suda, H., Haraya, K., 2005. Novel carbon molecular sieve membranes derived from poly(phenylene oxide) and its derivatives for gas separation. *Chemistry Letters*, 34, 958–959.

[38] Grainger, D., Hägg, M.-B., 2008. The recovery by carbon molecular sieve membranes of hydrogen transmitted in natural gas networks. *International Journal of Hydrogen Energy*, 33, 2379–2388.

[39] He, X., 2017. Techno-economic feasibility analysis on carbon membranes for hydrogen purification. *Separation and Purification Technology*, 186, 117–124.

[40] Itoh, N., Haraya, K., 2000. A carbon membrane reactor. *Catalysis Today*, 56, 103–111.

[41] Sznejer, G., Sheintuch, M., 2004. Application of a carbon membrane reactor for dehydrogenation reactions. *Chemical Engineering Science*, 59, 2013–2021.

[42] Zhang, X., Hu, H., Zhu, Y., Zhu, S., 2006. Methanol steam reforming to hydrogen in a carbon membrane reactor system. *Industrial & Engineering Chemistry Research*, 45, 7997–8001.

[43] Harale, A., Hwang, H.T., Liu, P.K.T., Sahimi, M., Tsotsis, T.T., 2007. Experimental studies of a hybrid adsorbent-membrane reactor (HAMR) system for hydrogen production. *Chemical Engineering Science*, 62, 4126–4137.

[44] Abdollahi, M., Yu, J, Liu, P.K.T., Ciora, R., Sahimi, M., Tsotsis, T.T., 2010. Hydrogen production from coal-derived syngas using a catalytic membrane reactor based process. *Journal of Membrane Science*, 363, 160–169.

[45] Parsley, D., Ciora Jr., R.J., Flowers, D.L., Laukaitaus, J., Chen, A., Liu, P.K.T., Yu, J., Sahimi, M., Bonsu, A., Tsotsis, T.T., 2014. Field evaluation of carbon molecular sieve membranes for the separation and purification of hydrogen from coal- and biomass-derived syngas. *Journal of Membrane Science*, 450, 81–92.

[46] Sheintuch, M., Efremenko, I., 2004. Analysis of a carbon membrane reactor: from atomistic simulations of single-file diffusion to reactor design. *Chemical Engineering Science*, 59, 4739–4746.

[47] Briceño, K., Iulianelli, A., Montané, D, Garcia-Valls, R., Basile, A., 2012b. Carbon molecular sieve membranes supported on non-modified ceramic tubes for hydrogen separation in membrane reactors. *International Journal of Hydrogen Energy*, 37, 13536–13544.

[48] Sá, S., Silva, H., Sousa, J.M., Mendes, A., 2009. Hydrogen production by methanol steam reforming in a membrane reactor: palladium vs. carbon molecular sieve membranes. *Journal of Membrane Science*, 339, 160–170.

8

Carbon Membranes for Microfiltration/ Ultrafiltration/Nanofiltration

Bing Zhang and Yonghong Wu

School of Petrochemical Engineering, Shenyang University of Technology

8.1 Introduction

Carbon membranes (CMs) are newly developed inorganic membrane materials for fluid mixture separation. In addition to excellent separation performance, CMs have the advantages of outstanding thermal stability and chemical inertness [1]. Owing to these comprehensively outstanding properties, CMs have become increasingly attractive for a large variety of fields, including chemical engineering, pharmaceutics, biologics, environmental protection, transportation and so on [2, 3].

In general, the membrane surface of CMs is dominated by hydrophobicity that is more beneficial for separation of non-polar gases than polar liquid mixtures, e.g., water. The rapid development of organic polymeric membranes for gas separation is the current mainstream. Therefore, most researchers concentrate on gas separation of CMs. Before the beginning of this century, prior to the discovery of novel carbons such as fullerene, carbon nanotubes, grapheme, graphyle, carbon dots and so on, the in-depth nature of carbon materials was not well understood [4]. As a matter of fact, the edges or defects in carbon materials have a large number of active chemical groups [5]. This finding provides more opportunities for a large variety of designs and applications based on the structure and properties of CMs [6].

All membrane materials are subject to a trade-off relationship between permeability and selectivity [7]. Consequently, people are keen to develop novel membrane materials by combining the easily tunable porous structure and the excellent separation efficiency of membranes such as carbon nanotube membranes and ordered porous membranes [8, 9]. However, the sub-nanoscale pore dimension of these membranes is too large for selective separation of gases, despite their use in removing large molecules, such as organic pollutants, from water. Although the effectiveness of this route for improving the separation property of CMs in comparison to traditional polymeric membranes has been demonstrated, the membranes should be called "carbon-doped polymeric membranes" rather than "carbon membranes" because the membrane matrix is mainly composed of polymer. Nevertheless, researchers have been inspired to explore functional carbon membranes with mesopores and macropores via hybridization or imitating ceramic membranes, or tailoring the surface properties, for the application of microfiltration, ultrafiltration and nanofiltration [10]. Undoubtedly, this exploration has broadened the horizon for the research and development of CMs.

Consequently, this chapter will focus on research into CMs for microfiltration, ultrafiltration and nanofiltration. There are many reports in the literature on the use of carbon nanotubes, activated carbon, graphene and carbon quantum dots as dopants into a continuous polymer matrix for liquid mixture separations such as microfiltration, ultrafiltration and nanofiltration. Researchers can find some well-documented reviews of those aspects [5, 11]. Here, only membranes composed from continuous carbon materials are summarized and reviewed in terms of preparation, modification and application.

8.2 General Fabrication of CMs

The production process of CMs mainly includes precursor selection, membrane formation, pyrolysis and modification.

8.2.1 Precursor Selection

CMs can be made from a large variety of carbonaceous materials, either natural or synthetic. Natural precursors include coal, pitch and biomass (e.g., lignin). In contrast to synthetic materials, the use of cheap mineral materials (e.g., coal, pitch), and in particular renewable biomass, has the advantage of low fabrication cost. However, in some cases, it is hard to control the microstructure and properties of the as-prepared CMs, to satisfactorily meet the separation requirement, because of the complicated and variable components in the starting products.

Synthetic precursors featuring uniform texture and stable properties are preferential for the fabrication of CMs without defects. Numerous synthetic materials have been reported to successfully form CMs, such as polyimide, phenolic resin, polyacrylonitrile, poly(furfuryl alcohol), polyvinylidene chloride, polyphenylene oxide, poly(phthalazinone ether sulfone ketone), polypyrrolone and so forth [12, 13].

Owing to differences in property, cost, availability and processability of raw materials, synthetic and natural materials are commonly applied to prepare CMs for gas separation and liquid separation, respectively.

8.2.2 Membrane Formation

Before pyrolysis, a defect-free thin organic layer is usually deposited on a permanent or temporary support to make supported or self-standing membranes, respectively. The supported CMs can be categorized into plate, tube, hollow fiber and spiral wound, whereas unsupported (or self-standing) CMs include plate, hollow fiber and capillary.

Unsupported plate CMs, which are facilely produced via casting or solvent evaporation, are restricted to the laboratory for academic study. Hollow fiber and capillary CMs have gained considerable attention owing to their huge specific volumetric membrane surface areas. Therefore, hollow fiber CMs have been commercially available for several years from Carbon Membranes Ltd (Israel) [14].

In practice, supported CMs are more promising for their robust mechanics to tolerate harsh environmental conditions in operation. To date, various coating technologies have been used in the preparation of CMs, such as brush coating, spray coating, dip-coating, drop-coating, ultrasonic spray coating, chemical vapor deposition, physical vapor deposition and so forth [13]. Supported plate CMs are commercially available from Blue Membranes GmbH (Germany) [15].

Prior to coating, porous supports should be carefully selected and pretreated; they include graphitic plate, carbon, ceramic, hollow fiber, stainless steel and polymeric substrates [16, 17]. In addition, nanoporous wafers made from glass, silicone and poly(tetrafluoroethylene) are frequently utilized as substrate to fabricate unsupported CMs during the solvent evaporation of the casting solution. The support should be selected to match the viscosity and nature of the membrane solution. Usually, supports with extremely large pore openings intake large amounts of membrane solution, which can devastate the formation of the surface layer. It is also detrimental for the surface layer to tightly adhere to the support, and for permeation of desired species if the pores are too small.

In some cases, CMs can be formed by directly compressing paste precursors into a membrane shape. This simple method is common for the production of CMs with a mesoporous structure for microfiltration and ultrafiltration, as will be elaborated in later sections.

8.2.3 Pyrolysis Conditions

Pyrolysis is a thermochemical decomposition process for organic precursors at elevated temperatures in the absence of an oxidative atmosphere to prevent excessive weight loss. It is the most conventional technique for preparing CMs by virtue of easy control of process parameters and facile operation with a temperature-programmable furnace. The fine manipulation of pyrolysis conditions is of course very important to achieve CMs with the desired structure and performance [12, 18]. The essential factors include pyrolysis temperature, atmosphere, heating rate and dwelling time [18, 19].

Pyrolysis temperature is usually set in the range of 500–1000 °C, and is one of the most vital parameters in determination of the pore texture and separation performance of CMs. Generally, the pore size and porosity of CMs first increase then decrease as the temperature is elevated from 500 to 1000 °C, as the result of the thermal degradation and rearrangement of precursor membranes [20].

Traditionally, two kinds of atmosphere can be selected during pyrolysis: inert flowing gas or vacuum [21]. A vacuum tends to yield more selective but less permeable CMs than an inert gas pyrolysis system [22, 23]. Sometimes, a trace amount of oxidative gas (e.g., O_2, CO_2, NH_3 or water steam) is used in purge gas with a concentration of 4–50 ppm to improve the porosity, pore size and permeance of CMs [24].

Heating rate determines the derivation rate of volatile component from precursor membranes during pyrolysis and consequently affects the formation of pores in CMs [25, 26]. Heating rate is generally set as 1–9 °C/min. Lower heating rates are preferable for forming small pores and high selectivity of CMs by increasing carbon crystallinity. The negative aspects of a slower heating rate are the time consumed and the high preparation costs [27].

Dwelling time can finely tune the microporous structure and the transport properties of CMs when the final pyrolysis temperature has been fixed. Extending dwelling time increases the selectivity and reduces the permeability of CMs due to the deepening of rearrangement reactions in the membrane matrix [28, 29].

8.2.4 Chemical Vapor Deposition

Chemical vapor deposition (CVD) can convert carbonaceous volatiles, such as methane, ethane, propane, ethylene, benzene and other hydrocarbons, or their mixture, into CMs in a short time on supports or substrates [30, 31]. A typical CVD is performed by exposing volatile precursors with the occurrence of thermal decomposition on the substrate surface to quickly transform a carbon layer. At the same time, the derived by-products are cleared from the system by purging gas flow through the reaction chamber to avoid any side effects. For this technology, the key operation factors include volatile precursors, deposition temperature, working pressure, mass flow rate, substrate size and precursor concentration in gas flow. [32]. The degree of crystallinity and regularity of CMs is expected to be reduced with increased deposition temperature [30].

In order to enhance the deposition efficiency, plasma or remote inductively coupled plasma are sometimes jointly coupled during the CVD process [33, 34]. High-energy ion bombardment can also enhance the gas separation performance, particularly the permeance of CMs [35].

8.2.5 Ion Irradiation

Ion irradiation can also achieve the thermal degradation of polymeric membranes so as to generate CMs. This technique has the merits of a short preparation period and low energy consumption, but is difficult to manipulate the process. Power and irradiation time are the key factors for determining the degree of pyrolysis and the variation of microstructure and separation performance of resultant CMs. Bombardment with high-energy ions alters the physical, chemical or electrical properties of the precursor surface so as to form CMs. However, improper operation might also destroy the structure and the separation performance due to energetic collision cascades. In the preparation of CMs, the operation factors include ion species (such as krypton, xenon, argon and helium), energy, irradiation time and frequency [36, 37]. By tailoring those process parameters, the thickness, pore size, density, length, shape and permeation property of CMs can be controlled well [38, 39].

In the following sections, some typical examples of CMs for the membrane filtration process for liquid mixtures will be introduced.

8.3 CMs for Microfiltration, Ultrafiltration and Nanofiltration

8.3.1 Compression Technology

This technology is usually adopted owing to its merit of easy operation and setup. Different from gas separation, it is only applicable for CMs with a rough structure, and for liquid separation.

Peydayesh et al. prepared amyloid fibril hybrid membranes on the surface of cellulose filters with a diameter of 25 mm by vacuum filtration of a homogenous aqueous mixture of 2% β-lactoglobulin fibril solution and 10 wt% activated carbon solution. The as-prepared macrofiltration CM has a surface area of 1.77 cm^2 [40].

Yang et al. formed a three-dimensional all-carbon nanofiltration membrane with a thickness of 4.26 μm, which consists of multi-walled carbon nanotubes interposed between graphene oxide nanosheets. The membrane is also abundant in two-dimensional nanochannels that can physically sieve antibiotic molecules through electrostatic interaction [41].

Boffa et al. fabricated composite graphene oxide–humic acid-like biopolymer membranes with a high degree of disorder, leading to the benefit of an increased water permeation rate [42].

Our group made microfiltration CMs by compressing a powder blend of phenolic resin as precursor, sodium carboxymethyl cellulose as binder and hexamethylenetetramine as curing agent. During preparation, fillers such as diatomite, polyacrylonitrile microspheres and so forth could be utilized to tailor the microstructure and its oil–water emulsion separation properties [43, 44].

8.3.2 Chemical Vapor Deposition

As elaborated earlier, CVD is also feasible in the preparation of CMs for liquid mixture separation. Lebedev et al. produced an ion-selective CM by CVD on nanofibers via filtration of a Nafen nanofiber suspension through a porous support followed by drying and sintering. The deposition had obviously decreased the pore size and porosity compared with the pristine ceramic support and provided outstanding ion selectivity.

Prihandana et al. deposited a diamond-like carbon layer on nanoporous poly(ether sulfones) membrane by ion-enhanced CVD. They investigated the effects of the composition of raw material (hexafluoroethane and acetylene) and deposition time. The adoption of fluoride in starting materials is favorable for lowering the hydrophilicity of resultant membranes [45].

Li et al. fabricated an 80-cm-long tubular-carbon-coated membrane by CVD of methane on a multilayered porous ceramic substrate at a temperature of 1000 °C. In comparison to the pristine carbon support, it has diminished pore size, as well as altered hydrophilicity, electric conductivity and heat conductivity. The membranes are expected to be applicable for microfiltration, ultrafiltration and nanofiltration [32].

Bae et al. dipped ceramic pipes in 0.5 M ferric sulfate solution, dried them at room temperature for 8 h, then heated under methane/nitrogen 20/80% (v/v) gas at 1100 °C to grow carbon whiskers by CVD of methane [46].

8.3.3 Membrane Coating and Pyrolysis

Pyrolysis is a general method for CM production from polymeric membranes although it is energy- and time-consuming. Kishore et al. synthesized CMs by pyrolyzing a phenol–formaldehyde resin film at 500 °C for 30 min in the absence of air and evaluated their physical properties and separation performance. The carbon membranes consist of a crystalline structure embedded in an amorphous carbon matrix [47]. Du et al. proposed a fabrication process for carbon nanofiber membranes by electrospinning a poly(acrylonitrile) solution followed by a preliminary heat treatment in the air at 250 °C and pyrolysis at 1000 °C in an inert atmosphere for 2 h [48].

The premise of pyrolysis is the formation of polymeric membranes on supports by various coating techniques, as will be reviewed in the following sections.

8.3.3.1 Spin-Coating

Spin-coating is performed by fixing the support on a central axis and then allowing the membrane-forming solution on the surface of the support to gradually spread into a film of uniform thickness under constant speed, driven by centrifugal force. According to the viscosity of the membrane-forming solution, the thickness and uniformity of the membrane layer can be controlled by adjusting the rotation speed and time.

Tseng et al. prepared superoleophilic and superhydrophobic CM by a spin-coating method to spread polyetherimide dope solution into a quartz filter paper, followed by drying and pyrolysis at 600 °C for 2h with a 5 °C/min ramping rate. The authors exploited the spin-coating parameters such as spin speed and interval times [49].

8.3.3.2 Cast-Coating

The method is usually feasible for a membrane-forming solution of relatively low dilution and viscosity when the complete membrane layer is allowed to form by natural spreading and consolidation.

Kayvani et al. prepared a hybrid ceramic/carbon membrane using a mixing-casting-sintering method. Briefly, Al_2O_3 powder was mixed with 0.02 M HNO_3 and highly porous activated carbon to form a 10% slurry. Then, it was ball-milled in the presence of Al_2O_3 balls and then air/gas removed from the solution. After that, the slurry was transferred to the disc-casting unit to produce a 1.5 mm thick green membrane, which was dried overnight and then sintered at 1150 °C under vacuum or Argon flow [50].

8.3.3.3 Dip-Coating

In this method, the support is immersed in a membrane-forming solution. After withdrawal from the solution, drying and carbonizing, a carbon membrane is formed, adhered to the support surface. The thickness and uniformity of the membrane layer can be adjusted by controlling the solution viscosity, soaking time and withdrawal speed.

Du et al. dipped a ceramic pipe into a polymer latex suspension, rotating at 600 rpm, for 60 s, then removed it from the suspension. Then, it was put in a quartz pipe, placed in an electric oven and heated under nitrogen atmosphere, maintaining the temperature at 300 °C to induce chloride desorption and holding at 1100 °C to stimulate polymer pyrolysis [48].

Derbel et al. elaborated an ultrafiltration carbon membrane prepared by deposition of a thin carbon film over a previously prepared microfiltration carbon membrane. The resulting ultrafiltration carbon membrane exhibits an average pore diameter of about 4 nm and a thickness of around 9 μm [51].

8.3.4 Sol–Gel Method

This method depends on the rule of solvent exchange from a sol precursor to form a gel layer on the surface of a support, and then drying and carbonizing to obtain a CM.

Qin et al. prepared mesoporous CMs on the outer surface of a fly-ash-based microfiltration ceramic membrane by the sol–gel method with resorcinol/formaldehyde. They investigated effects of the surfactant species and concentration, the resorcinol/formaldehyde concentration and the sol synthesis time. The pore size/distribution can be tailored to a specific separation process by adjusting the preparation conditions. These CMs are anticipated to be of use in ultrafiltration separation processes [52].

8.4 Modification of CMs

Hybridization with inorganic fillers is one of the most effective techniques to modify the molecular structure and property of existing precursors. There are two types of filler: porous and nonporous [53, 54].

Potential inorganic fillers include silica, zeolite, carbon molecular sieves, metal or salt, ordered mesoporous carbon, etc. [53, 55–58]. The addition of porous inorganic fillers can provide a large amount of additional porosity in the membrane matrix, while the nonporous ions, metals or metal oxides can improve the affinity of CMs for special gases. Depending on the function of fillers, the gas separation performance of CMs can be finely tuned by tailoring the dosage or types of fillers. The following section will give a more detailed description of different modification methods.

8.4.1 Surface Modification

Surface modification has two functions: manipulation of hydrophilicity/hydrophobicity and improving antifouling. The surface chemical groups are essential to the properties of CMs, and can drastically affect the adsorption coefficients and diffusivities of penetrants. Therefore, the separation process is subject to both the porous structure and the surface properties of CMs.

The surface properties include roughness, polarity and hydrophilicity (or hydrophobicity). Effective modification methods of surface groups are pre-oxidation, post oxidation, CVD and post pyrolysis. Pre-oxidation and post oxidation can introduce oxidative groups like hydroxyl moieties to the CM's surface, or form inter- or intra-molecule crosslinks. CVD and post-pyrolysis can reduce the surface functional groups by deepening the thermal reactions or carbonaceous deposition [59, 60].

In addition, the growth of carbon whiskers would inhibit the packing of particles into the microholes at the surface of the CM, reducing permeability. Analysis of the separation mechanisms proved that the packing density of the layer accumulated on the membrane reduces with the appearance of whiskers [46].

8.4.2 Pore Modification

Overall, there are four kinds of pore-modification strategy: pore forming, CVD, oxidation and templating.

Generally, the porous structure of CMs is very complicated, like that of activated carbon. Unlike crystalline materials, it is difficult to accurately understand because the structure of a CM has an amorphous, microporous architecture. From separation studies of penetrants, it appears that these materials have a wide range of pore size.

In order to acquire a satisfactory separation performance for CMs, the pore texture should meet the microstructure requirement. In most cases, the porous structure parameters must be fine-tuned after CMs are obtained or received as commercial products in the future, alongside the preconditioning of pore design and formation during pyrolysis. Based on the underlying rules of pore-structure formation, it is therefore essential to adjust the pore structure to the desired target structure. The pore structure can be adjusted by some post treatment methods, including CVD, post oxidation (or activation), post pyrolysis and templating.

Shah et al. developed a method for producing nanoporous CM on silica-modified stainless steel supports by pyrolysis of poly(furfuryl alcohol), with polyethylene glycol used as the pore-forming agent. The sub-micron-sized silica particles allow thin integral membranes to be formed after only two or three coats of the poly(furfuryl alcohol). Performance characteristics demonstrate that its separation property is comparable to that of commercial polymeric ultrafiltration membranes, with special stability even after prolonged exposure to 3 N NaOH [61]. They also conducted two regeneration protocols aiming at diminishing protein pollution: cleaning with 0.5 N NaOH at 50 °C and steam sterilization. After periodic back-pulsing, the water permeability is completely restored with no evidence of any structural alterations in the membrane.

Pugazhenthi et al. prepared a supported non-interpenetrating modified ultrafiltration CM by gas phase nitration using NO_x (a mixture of NO and NO_2) at 250 °C and subsequently aminated using hydrazine hydrate at 60 °C. Separation experiments on the chromic acid solution were carried out using unmodified (giving 96% rejection), nitrated (giving 84% rejection) and aminated (giving 88% rejection) CM [62].

Song et al. investigated the effect of pyrolysis conditions on microfiltration CMs by correlating the pore structure with the particle size distribution of raw coal powder [63]. They found that CMs made by inert gases have more "open" pores and a higher gas flux compared with those prepared under vacuum. The carbonization temperature also plays an important role in determining the pore structure and density of the CM. Additionally, a low heating rate favors a small pore size. Further microfiltration shows that the oil rejection coefficients of oily wastewater are up to 97%, with the oil concentrations of the permeate less than 10 mg/L, which meets the National Discharge Standard of China for wastewater [64].

8.4.3 Template Method

The template method is a very effective means of making pores in CMs. It uses organic components doped in the precursor membrane matrix that decompose or dissolve at a later stage to leave pores. Pores formed by this method will be distributed very evenly within the membrane matrix. The organic fillers are usually characteristic, with thermal labile segments or moieties such as polyvinylpyrrolidone [65], poly(ethylene glycol) [66], Pluronic triblock copolymer [67–69] and so on. After being thermally degraded they leave abundant porosity in the membrane matrix of CMs. In addition, inorganic dopants, such as silicon and porous carbon, can sometimes be used as templating agents. However, this kind of template needs to be removed by an additional etching step after membrane formation, to leave the pore structure.

Li et al. utilized an ethanol solution containing a phenolic resin and a Pluronic triblock copolymer to tailor commercialized polymeric hollow fiber ultrafiltration membranes. The surfactant self-assembled into the confined voids of the ultrafiltration membrane. After drying and pyrolysis, carbon hollow fiber membranes were obtained, the continuous membrane walls of which had an average thickness of 113 μm and a hierarchical pore structure: pores coming from hexagonally ordered mesoporous carbon had a pore diameter of ~4.3 nm and there were also disordered defect holes with a size of 8–50 nm randomly distributed inside the membrane matrix. The gas permeance results indicate that the membranes exhibit Knudsen diffusion behavior, confirming their good quality [70].

Strano et al. used spray deposition and pyrolysis of poly(furfuryl alcohol)/poly(ethylene glycol) mixtures on macroporous stainless steel supports. Poly(ethylene glycol) employed as a carbonization template creates a mesoporosity that leads to pores in the ultrafiltration range. When the molecular weight of poly(ethylene glycol) is below 2000 (g mol⁻¹), the template effect disappears for microfiltration of a polydisperse dextran solution, membrane film cracking occurs, and reproducibility is inferior [71].

8.4.4 Antifouling Ability

Antifouling ability is one of the biggest concerns for membrane materials in practice. The fouling mechanisms for liquid separation membranes in applications like microfiltration, ultrafiltration and nanofiltration can mainly be summarized as pore fouling and cake filtration [72]. It is believed that the service life of membrane materials is tightly bound to the contaminant medium, such as oil droplets, inorganic salts, bacteria and some solid or gel impurities [73]. In this regard, researchers endeavor to avoid the direct contact of such pollutants with the membrane surface and pore wall, in order to shield the membranes and extend their service time so as to strengthen the practicality and market competitiveness.

Focusing on these aspects, studies have been conducted on the modification of the membrane surface and the interfacial hydrophilicity or hydrophobicity by adjusting the surface chemical groups or growth of nano-whisker/nano-fibers, to prevent the deposition of pollutants on the membrane surface and inside pores [48]. Another factor is the introduction of external forces such as external disturbances or pulse/periodic flow disturbances, which interfere with the formation of the cake layer on the membrane surface during the separation process [74]. Pan et al. analyzed membrane-fouling mechanisms based on the modified Hermia's model. Under the operating conditions used in the study, there is a switch of fouling mechanisms from standard pore fouling to cake filtration at a small apparent flow rate and trans-membrane pressure difference. When the values of the two parameters are raised, the intermediate pore fouling becomes dominant after an early period of cake filtration [75].

8.5 Application Prospects

CMs are applicable for the separation and purification of liquid mixtures [76], including decolorization of coke furnace wastewater [77], pervaporation of azeotropic benzene–cyclohexane mixtures [78] and ultrafiltration of water solutions [62].

In comparison to polymeric membranes, CMs show major advantages in stability in concentrated acid/base solution and wide application fields [78], as will be introduced in later sections.

8.5.1 Membrane Filtration

The most traditional application of CM is as a filtration medium for wastewater treatment.

For industrial textile wastewater, it shows good performances in terms of permeate flux and efficiency (retention of chemical oxygen demand (COD) and salinity of 50% and 30%, respectively) and almost total retention of turbidity and color [79].

Du et al. prepared a poly-3-methylthiophene/CM with switchable surface wettability between superhydrophobicity and superhydrophilicity from coating poly-3-methylthiophene (P(3-MTH)) on the highly porous and electrically conductive carbon nanofiber membranes. The separation efficiency reaches 99.5% for surfactant-stabled emulsions with various oil–water ratios [48].

Ayadi et al. also gained a very high removal efficiency and retention (more than 99%) of oil from an oil-in-water emulsion stabilized by a surfactant through the adoption of CMs [80].

Derbel et al. identified that the hydrophobic characteristic of CMs allows the application of air gap membrane distillation for oily waste water treatment with high salt, giving oil retention of about 99% [51].

Peydayesh et al. removed heavy metals from aqueous solutions by bench-scale dead-end vacuum filtration setup of a CM. Platinum and silver were recovered from saturated membranes by high temperature thermal reduction. After filtration, saturated hybrid membranes were collected and burned for 1 h at temperatures above 2000 and 1000 °C for platinum and silver, respectively. The recovered metal nuggets were mechanically pressed to obtain thin films [40].

Yang et al. almost removed 99.23% tetracycline hydrochloride and 83.88% cationic dye methylene blue from water with a high water permeation of 16.12 L m^{-2} h^{-1} bar^{-1}, using all-carbon three-dimensional nanofiltration membranes [41].

Kishore et al. separated hexavalent chromium ions from an aqueous chromic acid solution, with apparent and intrinsic rejection of Cr^{6+} ions of the order of 70 and 90%, respectively [47].

Our group investigated the feasibility of CMs for removal of phosphorous acid, phenol and emulsified oil from wastewater: excellent separation properties and chemical stability were found during filtration [43, 44, 81].

8.5.2 Process Integration

In order to exert the comprehensive advantages of CM, some process integration has been attempted.

Li et al. coupled a coal-based CM with an electric field for enhanced oily wastewater treatment. They studied the effects of factors such as electric field strength, oil–water concentration, pH value, pump speed, electrolyte concentration and electrode distance, and found a significant increase in permeate water flux and oil rejection, as well as the antifouling ability of the membranes, after integration [82].

Pan et al. conducted experimental studies aiming at enhancing the microfiltration process through applying gas–liquid two-phase flow by injecting air into a titanium dioxide suspension during its microfiltration with the tubular porous carbon membrane as microfiltration medium. Results show that the gas–liquid two-phase flow can significantly enhance the microfiltration process and cause the maximum steady permeate flux to be 62% higher; the time required for reaching steady flux was accordingly shorter than that of the single-phase flow microfiltration process [83].

8.5.3 Membrane Bioreactor

A membrane bioreactor has been proven effective for wastewater and biomass purification.

Liu et al. developed a CM-aerated biofilm reactor to treat synthetic wastewater. The membrane exhibited a high degree of adhesion and good permeability. Continuous experiments showed that COD and NH^{4+}-N removal efficiency were 90 ± 2 and $92 \pm 4\%$ at removal rates of 35.6 ± 3.8 g COD/m^2 per day and 9.3 ± 0.6 g NH^{4+}-N/m^2 per day, respectively. Stoichiometric analysis revealed that 70–90% of the oxygen supplied was consumed by the nitrifier [84].

CMs were also tried for biomass purification by maintaining the activity of penetrants in the fields of protein ultrafiltration [61], biofuel separation through pervaporation [85], water removal from bioethanol [86] and oily wastewater treatment by microfiltration [64].

These promising applications offer a viable alternative to current membranes for separation and purification in advanced fields.

8.5.4 Energy Production

CMs are also applicable for fuel cell and electrocatalytic chemistry. Wang et al. [87] prepared CMs with excellent anticorrosion properties, hydrophobicity and conductivity, aiming at the modification of the stainless-steel electrode of a proton exchange membrane fuel cell in order to improve its work efficiency. Yang et al. [88] fabricated CMs from common filter paper, which exhibited excellent electrocatalytic activities toward the oxygen reduction reaction, high tolerance of methanol crossover and durability in alkaline solution.

8.6 Conclusions and Remarks

In summary, for CM preparation the key challenges lie in reducing the thickness of the membrane without introducing defects and in scaling up fabrication techniques to produce CMs with large surface areas. It is expected that one-atom-thick membranes with precise pore size and functionality can help in selective ion passage and salt rejection. Biological systems provide almost precise selectivity for the permeation of ions through protein channels: mimicking biological systems by controlling the features of the nanopores of artificial membranes can provide excellent selectivity for various ions. Graphene nanopores can be functionalized in such a way that the permeation barriers arising due to dehydration can be compensated by the interactions with the functional groups flanking the nanopores for better passage of ions [89]. Therefore, it is hoped that a well-designed and precise control can be prepared and applied.

Furthermore, CMs should be prepared in the form of a honeycomb or hollow fiber module to provide the additional benefits of a low drop in pressure and a high surface-to-volume ratio. It would also be advantageous to shift the thermodynamic equilibrium by improving the porous structure of CMs. In particular, it is necessary to find more process integrations and novel highly value-added areas from the viewpoint of promising applications for CMs.

References

[1] Q. Zhang, S. Chen, X. Quan, Y. Liu, H. Yu, H. Wang, Superpermeable nanoporous carbon-based catalytic membranes for electro-Fenton driven high-efficiency water treatment, *Journal of Materials Chemistry A*, 6 (2018) 23502–23512.

[2] T. Pietraß, Carbon-based membranes, *MRS Bulletin*, 31 (2011) 765–769.

[3] D.R. Paul, Creating new types of carbon-based membranes, *Science*, 335 (2012) 413.

[4] E.J. Leonhardt, R. Jasti, Emerging applications of carbon nanohoops, *Nature Reviews Chemistry*, 3 (2019) 672–686.

[5] M. Selvaraj, A. Hai, F. Banat, M.A. Haija, Application and prospects of carbon nanostructured materials in water treatment: A review, *Journal of Water Process Engineering*, 33 (2020) 100996.

[6] C. Sealy, Carbon-based membranes fill the gap, *Materials Today*, 19 (2016) 558.

[7] L.M. Robeson, The upper bound revisited, *Journal of Membrane Science*, 320 (2008) 390–400.

[8] B. Mi, Scaling up nanoporous graphene membranes, *Science*, 364 (2019) 1033.

[9] D.S. Sholl, J.K. Johnson, Making high-flux membranes with carbon nanotubes, *Science*, 312 (2006) 1003.

[10] Y. Wu, Y. Xia, X. Jing, P. Cai, A.D. Igalavithana, C. Tang, D.C.W. Tsang, Y.S. Ok, Recent advances in mitigating membrane biofouling using carbon-based materials, *Journal of Hazardous Materials*, 382 (2020) 120976.

[11] M. Bassyouni, M.H. Abdel-Aziz, M.S. Zoromba, S.M.S. Abdel-Hamid, E. Drioli, A review of polymeric nanocomposite membranes for water purification, *Journal of Industrial and Engineering Chemistry*, 73 (2019) 19–46.

[12] W.N.W. Salleh, A.F. Ismail, T. Matsuura, M.S. Abdullah, Precursor selection and process conditions in the preparation of carbon membrane for gas separation: A review, *Separation and Purification Technology*, 40 (2011) 261–311.

[13] S.M. Saufi, A.F. Ismail, Fabrication of carbon membranes for gas separation - A review, *Carbon*, 42 (2004) 241–259.

[14] S. Lagorsse, Carbon molecular sieve membranes Sorption, kinetic and structural characterization, *Journal of Membrane Science*, 241 (2004) 275–287.

[15] S. Lagorsse, A. Leite, F.D. Magalhães, N. Bischofberger, J. Rathenow, A. Mendes, Novel carbon molecular sieve honeycomb membrane module: configuration and membrane characterization, *Carbon*, 43 (2005) 809–819.

[16] J.O. Titiloye, I. Hussain, Synthesis and characterization of silicalite-1/carbon-graphite membranes, *Journal of Colloid and Interface Science*, 318 (2008) 50–58.

[17] C.W. Song, T.H. Wang, X.Y. Wang, J.S. Qiu, Y.M. Cao, Preparation and gas separation properties of poly(furfuryl alcohol)-based C/CMS composite membranes, *Separation and Purification Technology*, 58 (2008) 412–418.

[18] T.A. Centeno, J.L. Vilas, A.B. Fuertes, Effects of phenolic resin pyrolysis conditions on carbon membrane performance for gas separation, *Journal of Membrane Science*, 228 (2004) 45–54.

[19] K.M. Steel, W.J. Koros, An investigation of the effects of pyrolysis parameters on gas separation properties of carbon materials, *Carbon*, 43 (2005) 1843–1856.

[20] C.J. Anderson, S.J. Pas, G. Arora, S.E. Kentish, A.J. Hill, S.I. Sandler, G.W. Stevens, Effect of pyrolysis temperature and operating temperature on the performance of nanoporous carbon membranes, *Journal of Membrane Science*, 322 (2008) 19–27.

[21] J. Su, A.C. Lua, Effects of carbonisation atmosphere on the structural characteristics and transport properties of carbon membranes prepared from Kapton® polyimide, *Journal of Membrane Science*, 305 (2007) 263–270.

[22] M. Yamamoto, K. Kusakabe, J.-I. Hayashi, S. Morooka, Carbon molecular sieve membrane formed by oxidative carbonization of a copolyimide film coated on a porous support tube, *Journal of Membrane Science*, 133 (1997) 195–205.

[23] D.T. Clausi, W.J. Koros, Formation of defect-free polyimide hollow fiber membranes for gas separations, *Journal of Membrane Science*, 167 (2000) 79–89.

[24] M. Kiyono, P.J. Williams, W.J. Koros, Effect of pyrolysis atmosphere on separation performance of carbon molecular sieve membranes, *Journal of Membrane Science*, 359 (2010) 2–10.

[25] H. Suda, K. Haraya, Gas permeation through micropores of carbon molecular sieve membranes derived from Kapton polyimide, *Journal of Physical Chemistry B*, 101 (1997) 3988–3994.

[26] A.B. Fuertes, D.M. Nevskaia, T.A. Centeno, Carbon composite membranes from Matrimid (R) and Kapton (R) polyimides for gas separation, *Microporous and Mesoporous Materials*, 33 (1999) 115–125.

[27] W.N.W. Salleh, A.F. Ismail, Effects of carbonization heating rate on CO_2 separation of derived carbon membranes, *Separation and Purification Technology*, 88 (2012) 174–183.

[28] L.I.B. David, A.F. Ismail, Influence of the thermastabilization process and soak time during pyrolysis process on the polyacrylonitrile carbon membranes for O_2/N_2 separation, *Journal of Membrane Science*, 213 (2003) 285–291.

[29] B. Zhang, T.H. Wang, S.H. Zhang, J.S. Qiu, X.G. Han, Preparation and characterization of carbon membranes made from poly(phthalazinone ether sulfone ketone), *Carbon*, 44 (2006) 2764–2769.

[30] L.H. Lai, J.S. Yang, S.T. Shiue, Characteristics of carbon films prepared by thermal chemical vapor deposition using camphor, *Thin Solid Films*, 556 (2014) 544–551.

[31] Y. Ito, C. Christodoulou, M.V. Nardi, N. Koch, H. Sachdev, K. Mullen, Chemical vapor deposition of N-doped graphene and carbon films: the role of precursors and gas phase, *ACS Nano*, 8 (2014) 3337–3346.

[32] Y.Y. Li, T. Nomura, A. Sakoda, M. Suzuki, Fabrication of carbon coated ceramic membranes by pyrolysis of methane using a modified chemical vapor deposition apparatus, *Journal of Membrane Science*, 197 (2002) 23–35.

[33] L.H. Cheng, Y.J. Fu, K.S. Liao, J.T. Chen, C.C. Hu, W.S. Hung, K.R. Lee, J.Y. Lai, A high-permeance supported carbon molecular sieve membrane fabricated by plasma-enhanced chemical vapor deposition followed by carbonization for CO_2 capture, *Journal of Membrane Science*, 460 (2014) 1–8.

[34] L.J. Wang, F.C.N. Hong, Surface structure modification on the gas separation performance of carbon molecular sieve membranes, *Vacuum*, 78 (2005) 1–12.

[35] L.J. Wang, F.C.N. Hong, Carbon-based molecular sieve membranes for gas separation by inductively-coupled-plasma chemical vapor deposition, *Microporous and Mesoporous Materials*, 77 (2005) 167–174.

[36] Y. Koval, A. Geworski, K. Gieb, I. Lazareva, P. Muller, Fabrication and characterization of glassy carbon membranes, *Journal of Vacuum Science and Technology B*, 32 (2014). 042001.

[37] S. Hukushima, K. Sode, K. Yamazaki, Y. Suzuki, H. Kawakami, Carbon structure in polyimide membrane formed by ion irradiation, *Journal of Photopolymer Science and Technology*, 23 (2010) 507–510.

[38] Z. Laušević, P.Y. Apel, I.V. Blonskaya, The production of porous glassy carbon membranes from swift heavy ion irradiated Kapton, *Carbon*, 49 (2011) 4948–4952.

[39] M. Iwase, A. Sannomiya, S. Nagaoka, Y. Suzuki, M. Iwaki, H. Kawakami, Gas permeation properties of asymmetric polyimide membranes with partially carbonized skin layer, *Macromolecules*, 37 (2004) 6892–6897.

[40] M. Peydayesh, S. Bolisetty, T. Mohammadi, R. Mezzenga, Assessing the binding performance of amyloid–carbon membranes toward heavy metal ions, *Langmuir*, 35 (2019) 4161–4170.

[41] G.-h. Yang, D.-d. Bao, D.-q. Zhang, C. Wang, L.-l. Qu, H.-t. Li, Removal of antibiotics from water with an all-carbon 3D nanofiltration membrane, *Nanoscale Research Letters*, 13 (2018) 146.

[42] V. Boffa, H. Etmimi, P.E. Mallon, H.Z. Tao, G. Magnacca, Y.Z. Yue, Carbon-based building blocks for alcohol dehydration membranes with disorder-enhanced water permeability, *Carbon*, 118 (2017) 458–466.

[43] X. Hong, B. Zhang, X. Zhang, Y. Wu, T. Wang, J. Qiu, Tailoring the structure and property of microfiltration carbon membranes by polyacrylonitrile-based microspheres for oil-water emulsion separation, *Journal of Water Process Engineering*, 32 (2019) 100973.

[44] X. Zhang, B. Zhang, Y. Wu, T. Wang, J. Qiu, Preparation and characterization of a diatomite hybrid microfiltration carbon membrane for oily wastewater treatment, *Journal of the Taiwan Institute of Chemical Engineers*, 89 (2018) 39–48.

[45] S.G. Prihandana, I. Sanada, H. Ito, M. Noborisaka, Y. Kanno, T. Suzuki, N. Miki, Antithrombogenicity of fluorinated diamond-like carbon films coated nano porous polyethersulfone (PES) membrane, *Materials*, 6 (2013) 309-432.

[46] S.-D. Bae, M. Sagehashi, A. Sakoda, Prevention of microparticle blocking in activated carbon membrane filtration with carbon whisker, *Journal of Membrane Science*, 252 (2005) 155–163.

[47] N. Kishore, S. Sachan, K.N. Rai, A. Kumar, Synthesis and characterization of a nanofiltration carbon membrane derived from phenol–formaldehyde resin, *Carbon*, 41 (2003) 2961–2972.

[48] L. Du, X. Quan, X. Fan, S. Chen, H. Yu, Electro-responsive carbon membranes with reversible superhydrophobicity/superhydrophilicity switch for efficient oil/water separation, *Separation and Purification Technology*, 210 (2019) 891–899.

[49] H.-H. Tseng, J.-C. Wu, Y.-C. Lin, G.-L. Zhuang, Superoleophilic and superhydrophobic carbon membranes for high quantity and quality separation of trace water-in-oil emulsions, *Journal of Membrane Science*, 559 (2018) 148–158.

[50] A. Kayvani Fard, A. Bukenhoudt, M. Jacobs, G. McKay, M.A. Atieh, Novel hybrid ceramic/carbon membrane for oil removal, *Journal of Membrane Science*, 559 (2018) 42–53.

[51] I. Derbel, R. Ben Amar, Preparation of new tubular carbon ultrafiltration membrane for oily wastewater treatment by air gap membrane distillation, *Desalination and Water Treatment*, 124 (2018) 21–29.

[52] G. Qin, H. Wen, W. Wei, Effects of processing parameters on the pore structure of mesoporous carbon membranes, *Desalination and Water Treatment*, 57 (2016) 23536–23545.

[53] H.H. Tseng, P.T. Shiu, Y.S. Lin, Effect of mesoporous silica modification on the structure of hybrid carbon membrane for hydrogen separation, *International Journal of Hydrogen Energy*, 36 (2011) 15352–15363.

[54] Y.C. Xiao, M.L. Chng, T.S. Chung, M. Toriida, S. Tamai, H.M. Chen, Y.C.J. Jean, Asymmetric structure and enhanced gas separation performance induced by in situ growth of silver nanoparticles in carbon membranes, *Carbon*, 48 (2010) 408–416.

[55] M. Teixeira, M. Campo, D.A. Tanaka, M.A. Tanco, C. Magen, A. Mendes, Carbon-Al_2O_3-Ag composite molecular sieve membranes for gas separation, *Chemical Engineering Research and Design*, 90 (2012) 2338–2345.

[56] X.Y. Zhang, H.Q. Hu, Y.D. Zhu, S.W. Zhu, Effect of carbon molecular sieve on phenol formaldehyde novolac resin based carbon membranes, *Separation and Purification Technology*, 52 (2006) 261–265.

[57] B. Zhang, Y. Shi, Y.H. Wu, T.H. Wang, J.S. Qiu, Towards the preparation of ordered mesoporous carbon/carbon composite membranes for gas separation, *Separation Science and Technology*, 49 (2014) 171–178.

[58] Q.L. Liu, T.H. Wang, C.H. Liang, B. Zhang, S.L. Liu, Y.M. Cao, J.S. Qiu, Zeolite married to carbon: A new family of membrane materials with excellent gas separation performance, *Chemistry of Materials*, 18 (2006) 6283–6288.

[59] A. Stein, Z.Y. Wang, M.A. Fierke, Functionalization of porous carbon materials with designed pore architecture, *Advanced Materials*, 21 (2009) 265–293.

[60] S.M. Mahurin, J.S. Lee, X. Wang, S. Dai, Ammonia-activated mesoporous carbon membranes for gas separations, *Journal of Membrane Science*, 368 (2011) 41–47.

[61] T.N. Shah, H.C. Foley, A.L. Zydney, Development and characterization of nanoporous carbon membranes for protein ultrafiltration, *Journal of Membrane Science*, 295 (2007) 40–49.

[62] G. Pugazhenthi, S. Sachan, N. Kishore, A. Kumar, Separation of chromium (VI) using modified ultrafiltration charged carbon membrane and its mathematical modeling, *Journal of Membrane Science*, 254 (2005) 229–239.

[63] C. Song, T. Wang, J. Qiu, Pore structure prediction of coal-based microfiltration carbon membranes, *Chinese Science Bulletin*, 55 (2010) 1325–1330.

[64] C. Song, T. Wang, Y. Pan, J. Qiu, Preparation of coal-based microfiltration carbon membrane and application in oily wastewater treatment, *Separation and Purification Technology*, 51 (2006) 80–84.

[65] W.N.W. Salleh, A.F. Ismail, Effect of stabilization condition on PEI/PVP-based carbon hollow fiber membranes properties, *Separation Science and Technology*, 48 (2013) 1030–1039.

[66] W.Z. Wan Nurul Huda, M.A. Ahmad, A comparison of carbon molecular sieve (CMS) membranes with polymer blend CMS membranes for gas permeation applications, *ASEAN Journal of Chemical Engineering*, 12 (2012) 51–58.

[67] S. Tanaka, N. Nakatani, A. Doi, Y. Miyake, Preparation of ordered mesoporous carbon membranes by a soft-templating method, *Carbon*, 49 (2011) 3184–3189.

[68] X. Zhao, W. Li, S.X. Liu, Ordered mesoporous carbon membrane prepared from liquefied larch by a soft method, *Materials Letters*, 126 (2014) 174–177.

[69] B. Zhang, Y. Shi, Y. Wu, T. Wang, J. Qiu, Preparation and characterization of supported ordered nanoporous carbon membranes for gas separation, Journal of Applied Polymer Science, 131 (2014) app.39925.

[70] J. Li, J. Qi, C. Liu, L. Zhou, H. Song, C. Yu, J. Shen, X. Sun, L. Wang, Fabrication of ordered mesoporous carbon hollow fiber membranes via a confined soft templating approach, *Journal of Materials Chemistry A*, 2 (2014) 4144–4149.

[71] M.S. Strano, A.L. Zydney, H. Barth, G. Wooler, H. Agarwal, H.C. Foley, Ultrafiltration membrane synthesis by nanoscale templating of porous carbon, *Journal of Membrane Science*, 198 (2002) 173–186.

[72] X. Xu, J. Li, N. Xu, Y. Hou, J. Lin, Visualization of fouling and diffusion behaviors during hollow fiber microfiltration of oily wastewater by ultrasonic reflectometry and wavelet analysis, *Journal of Membrane Science*, 341 (2009) 195–202.

[73] A. Salahi, A. Gheshlaghi, T. Mohammadi, S.S. Madaeni, Experimental performance evaluation of polymeric membranes for treatment of an industrial oily wastewater, *Desalination*, 262 (2010) 235–242.

[74] D. Li, C. Li, Z. Gao, Q. Jin, On active disturbance rejection in temperature regulation of the proton exchange membrane fuel cells, *Journal of Power Sources*, 283 (2015) 452–463.

[75] Y. Pan, W. Wang, T. Wang, P. Yao, Fabrication of carbon membrane and microfiltration of oil-in-water emulsion: An investigation on fouling mechanisms, *Separation and Purification Technology*, 57 (2007) 388–393.

[76] A.S. Damle, S.K. Gangwal, V.K. Venkataraman, Carbon membranes for gas separation: Developmental studies, *Gas Separation and Purification*, 8 (1994) 137–147.

[77] A. Sakoda, T. Nomura, M. Suzuki, Activated carbon membrane for water treatments: Application to decolorization of coke furnace wastewater, *Adsorption*, 3 (1997) 93–98.

[78] Y. Sakata, A. Muto, M.A. Uddin, H. Suga, Preparation of porous carbon membrane plates for pervaporation separation applications, *Separation and Purification Technology*, 17 (1999) 97–100.

[79] N. Tahri, I. Jedidi, S. Cerneaux, M. Cretin, R. Ben Amar, Development of an asymmetric carbon microfiltration membrane: Application to the treatment of industrial textile wastewater, *Separation and Purification Technology*, 118 (2013) 179–187.

[80] S. Ayadi, I. Jedidi, S. Lacour, S. Cerneaux, M. Cretin, R.B. Amar, Preparation and characterization of carbon microfiltration membrane applied to the treatment of textile industry effluents, *Separation Science and Technology*, 51 (2016) 1022–1029.

[81] W. Zhao, Y. Liang, Y. Wu, D. Wang, B. Zhang, Removal of phenol and phosphoric acid from wastewater by microfiltration carbon membranes, *Chemical Engineering Communications*, 205 (2018) 1432–1441.

[82] C. Li, C. Song, P. Tao, M. Sun, Z. Pan, T. Wang, M. Shao, Enhanced separation performance of coal-based carbon membranes coupled with an electric field for oily wastewater treatment, *Separation and Purification Technology*, 168 (2016) 47–56.

[83] Y.Q. Pan, F.Y. Li, W. Zhang, T.T. Wang, T.H. Wang, Application of gas-liquid two-phase flow in microfiltration of titanium dioxide suspension through tubular porous carbon membrane, Gao Xiao Hua Xue Gong Cheng Xue Bao/*Journal of Chemical Engineering of Chinese Universities*, 23 (2009) 34–38.

[84] H. Liu, F. Yang, T. Wang, Q. Liu, S. Hu, Carbon membrane-aerated biofilm reactor for synthetic wastewater treatment, *Bioprocess and Biosystems Engineering*, 30 (2007) 217–224.

[85] P.S. Tin, H.Y. Lin, R.C. Ong, T.S. Chung, Carbon molecular sieve membranes for biofuel separation, *Carbon*, 49 (2011) 369–375.

[86] K.S. Liao, Y.J. Fu, C.C. Hu, J.T. Chen, D.W. Lin, K.R. Lee, K.L. Tung, Y.C. Jean, J.Y. Lai, Microstructure of carbon molecular sieve membranes and their application to separation of aqueous bioethanol, *Carbon*, 50 (2012) 4220–4227.

[87] T. Wang, C.X. Zhang, X. Sun, Y.X. Guo, H. Guo, J. Tang, H.R. Xue, M.Z. Liu, X.X. Zhang, L. Zhu, Q.Q. Xie, J.P. He, Synthesis of ordered mesoporous boron-containing carbon films and their corrosion behavior in simulated proton exchange membrane fuel cells environment, *Journal of Power Sources*, 212 (2012) 1–12.

[88] W. Yang, Y. Zhai, X. Yue, Y. Wang, J. Jia, From filter paper to porous carbon composite membrane oxygen reduction catalyst, *Chemical Communications*, 50 (2014) 11151–11153.

[89] C. Owais, A. James, C. John, R. Dhali, R.S. Swathi, Selective permeation through one-atom-thick nanoporous carbon membranes: Theory reveals excellent design strategies! *The Journal of Physical Chemistry B*, 122 (2018) 5127–5146.

9

Carbon Membranes for Other Applications

Sergio San Martin Gomez[a,b], Izumi Kumakiri[a] and Xuezhong He[c,d]

[a]Graduate School of Sciences and Technology for Innovation, Yamaguchi University

[b]Department of Chemical and Biomolecular Engineering, ETSIIyT, University of Cantabria

[c]Department of Chemical Engineering, Norwegian University of Science and Technology

[d]Department of Chemical Engineering, Guangdong Technion Israel Institute of Technology (GTIIT)

9.1 Introduction

Besides the applications described in Chapters 5–8, several other industrial applications have potential for carbon membranes. This chapter discusses other applications of carbon membranes for gas separations and liquid separations such as CO_2 capture from flue gas, olefin/paraffin separation in petrochemical industry, and carbon membranes in an integrated photosynthetic process.

9.2 Carbon Membranes for CO_2 Capture from Flue Gas

The International Energy Outlook 2010 (IEO2010) reference case predicted that global energy-related carbon dioxide (CO_2) emissions will increase from 29.7 billion metric tons in 2007 to 33.8 billion metric tons in 2020 and 42.4 billion metric tons in 2035. The control of anthropogenic emissions of greenhouse gases is one of the most challenging environmental issues owing to the implications for global climate change. CO_2 capture and storage is becoming one of the most attractive technologies to reduce CO_2 emissions, as it will allow us to continue to use fossil fuels but without causing significant CO_2 emissions.

Different technologies such as chemical and physical absorption, membrane separation, physical adsorption, cryogenic distillation, and chemical looping can be used for CO_2 capture [1]. Great efforts have recently been made toward CO_2 capture using gas separation membranes, and examples are found in the literature [2–12]. Some polymeric membranes have demonstrated a high separation performance at pilot scale [13]. Membrane stability may decrease when exposed to SO_x and NO_x, which usually exist in flue gas, thus shortening membrane lifetime, which will increase the capital cost. The potential of carbon membranes for CO_2 capture has been reported by He et al. [11, 12]. The reported membrane performance was quite low: a specific capture cost of 136 $/ton CO_2 avoided was theoretically estimated based on HYSYS simulation [11]. Improving membrane performance (especially a 2.5 × CO_2 permeance) has the potential to reduce the cost to 57 $/ton CO_2 avoided.

9.3 Carbon Membranes for Olefin/Paraffin Separation

In petrochemical plants, there are numerous gas streams containing valuable components that can be recovered and separated. These are typically non-reacted monomers, by-products from reactors, inert substances, solvents, and carrier gas. Membrane separation shows great advantages for olefin/paraffin separation due to its low energy consumption and continuous operation [14]. There is potential for carbon membranes to be used for gas separation in this field.

A case study on the separation of alkanes and alkenes was performed by Hägg et al. [15]. Their systems separated propane–propene and propane–ethene. As the alkanes and alkenes are chemically and physically quite similar with almost identical critical properties, they must be separated on the basis of molecular size: propene and propane have Lennard-Jones diameters of 4.7 Å and 5.1 Å, respectively. There are few competing technologies in this market: perhaps only distillation, which is highly energy demanding. A carefully tailored carbon membrane can separate these two components according to the molecular sieving mechanism. Lab-scale carbon hollow fiber modules produced by Carbon Membranes Ltd. were tested by mixed gas permeation with propene/propane at 50 °C, at feed pressure 3 bar and permeate pressure 1 bar. The permeance of propene was 0.07 m³ (STP)/(m² h bar) (i.e., 27 GPU), with a selectivity of 50. It was also reported that it is favorable to operate carbon membrane systems at a higher temperature.

Xu et al. [16] developed asymmetric carbon molecular sieve (CMS) hollow fiber membranes and advanced processes for olefin/paraffin separation. CMS membranes made from polyimide promise to exceed the performance upper bound of polymer materials and have demonstrated good stability for olefin–paraffin separation. High selectivities for C_2H_4/C_2H_6 and C_3H_6/C_3H_8 of 4 and 21 were reported, which are much higher than those of most polymeric membranes. Moreover, carbon membrane separation processes based on the olefin-selective feature have even more potential applications than the previous retrofitting concepts that were based on de-bottlenecking a C2-splitter, C3-splitter, de-ethanizer, etc.

Xu et al. [17] carbonized polyimide membranes and performed mixed gas tests with ethene/ethane at 35 °C, feed pressure 8 bar and permeate pressure 1 bar. After about 40 hours the ethene permeance stabilized at 13 GPU and a selectivity of 4 was obtained. Therefore, significant energy savings and a reduced footprint can be achieved by reconfiguring the separations in the hydrocarbon processes, especially applying carbon membranes.

9.4 Carbon Membranes for Artificial Photosynthesis

Artificial photosynthesis is a process that uses sunlight to produce chemicals from water and CO_2 separated from flue gas, for example. There are various ideas for using the solar energy efficiently and many projects are ongoing, such as An Artificial Leaf: a photo-electro-catalytic cell from earth-abundant materials for sustainable solar production of CO_2-based chemicals and fuels (A-LEAF, 2017–2020, European project Horizon 2020) [18], the soap film based artificial photosynthesis project (SoFiA, 2019–2022, European project Horizon 2020) [19], and the Rheticus project (Evonic, 2018–) [20].

In the artificial photosynthetic chemical process project (ARPChem, 2012–2022, New Energy and Industrial Technology Development Organization, Japan), a membrane-integrated system is under development, and a carbon membrane is one of the potential candidates [21]. Figure 9.1 shows a schematic image of the concept. After photocatalytic water splitting, a membrane is employed to separate hydrogen from oxygen and water vapor. Then, hydrogen and CO_2 are converted to olefins (Figure 9.1). Hydrogen can be produced by electrolysis using electricity produced using, for example, photovoltaic (PV) cells (Figure 9.1). The configuration of a membrane-integrated hydrogen production process is simpler than an electrolysis-based process, which may reduce the capital cost and is an advantage for recycling after the lifetime of the system.

Let us compare these two processes in terms of the photosynthesis area needed to produce 100 Nm³ of hydrogen per hour. Table 9.1 shows the energy requirements and hydrogen production with the PV cell-electrolysis case obtained from an industrial facility of Energiepark Mainz, the largest global power-to-gas plant with polymer electrolyte membrane electrolysis [22].

The energy required to produce 100 Nm³/h of hydrogen will be:

$$100 \, \text{Nm}^3\!\!\Big/\!\!_h \text{ of } H_2 \times \frac{432.5 \, \text{MWh}}{533.22 \, \text{Nm}^3\!\!\Big/\!\!_h} \times \frac{1}{146 \, \text{working hours}}$$

$$= 0.55 \, \text{MW} = 555.5 \, \text{kW}$$

(9.1)

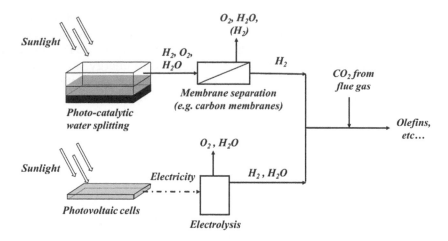

FIGURE 9.1
A schematic illustration of carbon membrane application in an artificial photosynthetic process.

TABLE 9.1

EnergiePark Mainz technical data [22].

Energy requirements (MWh)	Hydrogen production (Nm³/h)
432.5	533.22

Under the best possible conditions, a single industrial solar panel can produce up to 250 W [23]. Then, the minimum number of panels to produce
hydrogen at industrial level is 2,223 with a total panel area of ca. 3,680 m².

In contrast, photocatalytic water splitting by photocatalysis is still at bench
scale. Using the hydrogen production rate under solar light as $1.016 \, L \cdot m^{-2} \cdot h^{-1}$
(506 mL collected in 30 minutes using a 1 m² panel) [24], the total panel area
to obtain 100 Nm³/h hydrogen will be:

$$100 \, \text{Nm}^3\!\Big/\!_\text{h} \, of \, H_2 * \frac{1 \text{h} \cdot \text{m}^2}{0.1016 \, \text{Nm}^3} = 98425.2 \approx 98,400 \, \text{m}^2 \qquad (9.2)$$

As membrane separation uses pressure difference across the membrane
as a driving force, it is not possible to remove all the hydrogen produced.
Assuming a recovery rate (the amount of hydrogen separated by a membrane
divided by the produced mass) of 90%, the required photocatalytic panel
area is 108,000 m². If the photocatalysis is improved to e.g. $37.3 \, L \cdot m^{-2} \cdot h^{-1}$,
which is the value reported with UV light [24], the required area is decreased
to ca. 3,000 m². In this case, the required photosynthesis area is smaller than

that of the PV-cell electrolysis case. Accordingly, the development of photo-catalysis is one of the key technologies for the realization of a membrane-integrated system.

Another key development is the separation process. CMS membranes are a promising candidate for hydrogen separation at ambient temperature. However, water vapor can adsorb to the membrane and reduces its perfor-mance [11, 25]. Therefore, preventing the adverse influence of water and improving the hydrogen flux are necessary before carbon membranes can be used.

9.5 Carbon Membranes for Organic Solvent Separation

The pore size of carbon membranes can be adjusted by tuning the carbon-ization conditions and employment of proper post treatment as described in Chapter 2, and thus the membranes are good candidates for the separa-tion of organic molecules due to their stability and the avoidance of expen-sive supports or complex multi-step fabrication processes [26]. However, the critical challenge is the creation of "mid-range" (5–9 Å) microstructures that allow for facile permeation of organic solvents and selection between simi-larly sized guest molecules. A carbon membrane made from a microporous polymer (PIM-1) with an average pore size of 5.1 Å under low concentra-tions of hydrogen gas has been reported to show a high p-xylene/o-xylene selectivity of 14.7 for equimolar mixture tests [26]. The reported approach was successfully extended to hollow fiber membranes operating in organic solvent reverse-osmosis mode, highlighting the potential of this method to be translated from the laboratory to the field.

References

[1] He, X.; Yu, Q.; Hägg, M.-B., CO_2 capture. In *Encyclopedia of Membrane Science and Technology*, Hoek, E. M. V.; Tarabara, V. V., Eds. John Wiley & Sons, Inc., 2013.

[2] Carapellucci, R.; Milazzo, A., Membrane systems for CO_2 capture and their inte-gration with gas turbine plants. *Proc. Inst. Mech. Eng. Part A J. Power Energy* 2003, 217 (5), 505–517.

[3] Yang, H.; Xu, Z.; Fan, M.; Gupta, R.; Slimane, R. B.; Bland, A. E.; Wright, I., Progress in carbon dioxide separation and capture: A review. *J. Environ. Sci.* 2008, 20, 14–27.

[4] Bredesen, R.; Jordal, K.; Bolland, O., High-temperature membranes in power generation with CO_2 capture. *Chem. Eng. Process.* 2004, 43 (9), 1129–1158.

[5] Hagg, M. B.; Lindbrathen, A., CO$_2$ capture from natural gas fired power plants by using membrane technology. *Ind. Eng. Chem. Res.* 2005, 44 (20), 7668–7675.

[6] Huang, J.; Zou, J.; Ho, W. S. W., Carbon dioxide capture using a CO$_2$-selective facilitated transport membrane. *Ind. Eng. Chem. Res.* 2008, 47 (4), 1261–1267.

[7] Reijerkerk, S. R. Polyether based block copolymer membranes for CO$_2$ separation [PhD]. University of Twente, Enschede, 2010.

[8] Lin, H.; Freeman, B. D., Materials selection guidelines for membranes that remove CO$_2$ from gas mixtures. *J. Mol. Struct.* 2005, 739 (1–3), 57–74.

[9] Deng, L.; Kim, T.-J.; Hägg, M.-B., Facilitated transport of CO$_2$ in novel PVAm/PVA blend membrane. *J. Membr. Sci.* 2009, 340 (1–2), 154–163.

[10] Sandru, M.; Haukebø, S. H.; Hägg, M.-B., Composite hollow fiber membranes for CO$_2$ capture. *J. Membr. Sci.* 2010, 346 (1), 172–186.

[11] He, X.; Hägg, M.-B., Hollow fiber carbon membranes: Investigations for CO$_2$ capture. *J. Membr. Sci.* 2011, 378 (1–2), 1–9.

[12] He, X.; Hägg, M.-B., Hollow fiber carbon membranes: From material to application. *Chem. Eng. J.* 2013, 215–216 (0), 440–448.

[13] He, X., A review of material development in the field of carbon capture and the application of membrane-based processes in power plants and energy-intensive industries. *Energy Sust. Soc.* 2018, 8 (1), 34.

[14] Hou, J.; Liu, P.; Jiang, M.; Yu, L.; Li, L.; Tang, Z., Olefin/paraffin separation through membranes: from mechanisms to critical materials. *J. Mater. Chem. A* 2019, 7 (41), 23489–23511.

[15] Hagg, M.-B.; Lie, J. A.; Lindbrathen, A., Carbon molecular sieve membranes. A promising alternative for selected industrial applications. *Ann. N.Y. Acad. Sci.* 2003, 984 (1), 329–345.

[16] Xu, L.; Rungta, M.; Brayden, M. K.; Martinez, M. V.; Stears, B. A.; Barbay, G. A.; Koros, W. J., Olefins-selective asymmetric carbon molecular sieve hollow fiber membranes for hybrid membrane-distillation processes for olefin/paraffin separations. *J. Membr. Sci.* 2012, 423–424, 314–323.

[17] Xu, L.; Rungta, M.; Hessler, J.; Qiu, W.; Brayden, M.; Martinez, M.; Barbay, G.; Koros, W. J., Physical aging in carbon molecular sieve membranes. *Carbon* 2014, 80, 155–166.

[18] An Artificial Leaf: a photo-electro-catalytic cell from earth-abundant materials for sustainable solar production of CO$_2$-based chemicals and fuels, https://cordis.europa.eu/project/id/732840 (accessed May 3).

[19] Soap film based artificial photosynthesis, https://cordis.europa.eu/project/id/828838 (accessed May 3).

[20] Artificial photosynthesis: A contribution to the energy transition, https://www.creavis.com/sites/creavis/en/activities/current-projects/rheticus/ (accessed May 3).

[21] Yamada, T.; Domen, K., Development of sunlight driven water splitting devices towards future artificial photosynthetic industry. *ChemEngineering* 2018, 2 (3), 36.

[22] Kopp, M.; Coleman, D.; Stiller, C.; Scheffer, K.; Aichinger, J.; Scheppat, B., Energiepark Mainz: Technical and economic analysis of the worldwide largest power-to-gas plant with PEM electrolysis. *Int. J. Hydrogen Energy* 2017, 42 (19), 13311–13320.

[23] *Mitsubishi Electr.*, MLU series photovoltaic modules, n.d., https://www.mitsubishielectricsolar.com/images/uploads/documents/specs/MLU_spec_sheet_250W_255W.pdf (accessed May 3).

[24] Goto, Y.; Hisatomi, T.; Wang, Q.; Higashi, T.; Ishikiriyama, K.; Maeda, T.; Sakata, Y.; Okunaka, S.; Tokudome, H.; Katayama, M.; Akiyama, S.; Nishiyama, H.; Inoue, Y.; Takewaki, T.; Setoyama, T.; Minegishi, T.; Takata, T.; Yamada, T.; Domen, K., A particulate photocatalyst water-splitting panel for large-scale solar hydrogen generation. *Joule* 2018, 2 (3), 509–520.

[25] Jones, C. W.; Koros, W. J., Carbon composite membranes: A solution to adverse humidity effects. *Ind. Eng. Chem. Res.* 1995, 34 (1), 164–167.

[26] Ma, Y.; Jue, M. L.; Zhang, F.; Mathias, R.; Jang, H. Y.; Lively, R. P., Creation of well-defined "mid-sized" micropores in carbon molecular sieve membranes. *Angew. Chem. Int. Ed.* 2019, 58 (38), 13259–13265.

10

Future Perspectives

Xuezhong He

Department of Chemical Engineering, Guangdong Technion Israel Institute of Technology (GTIIT)
The Wolfson Department of Chemical Engineering, Technion - Israel Institute of Technology

This chapter discusses the future for carbon membrane development with respect to precursor selection and advanced carbonization processes to achieve high-performance carbon membranes at a low cost. Even though different precursors can be used to prepare carbon membranes, the most promising precursors, polyimide and cellulose, balance production cost and separation performance. The current carbon membranes perform well for CO_2/CH_4, CO_2/N_2, and H_2/CO_2 separation, but none of them have been successfully brought to the market at an industrial scale. A sustainable production process from precursors to carbon membranes should be pursued; technological advances in carbon membrane development should focus on asymmetric carbon hollow fiber membranes with no need for costly pretreatment. Renewable cellulose polymers using ionic liquids (ILs) as solvent are of particular interest. A new methodology combines experiment, chemometrics, and molecular modeling to develop cellulose-based carbon membranes, as shown in Figure 10.1. Screening ILs and cellulose feedstocks will allow the production of spin-defect-free, asymmetric cellulose hollow fibers to be fine-tuned, producing desirable structures and properties (e.g., porosity, degree of polymerization, crystallinity, hydrophilicity, etc.)The principles of non-equilibrium thermodynamics (cellulose/IL/H_2O ternary phase diagram) should be used to guide the spinning of cellulose hollow fibers from cellulose/IL systems.

Conventional carbonization processes using vacuum or inert N_2 gas suffer from low gas permeance and defect formation due to the accumulation/sintering of residual carbon ashes on the carbon surface/matrix. Thus, a novel carbonization process (see Figure 10.2) has been developed to prepare straight, mechanically strong, asymmetric carbon hollow fibers with high porosity and fewer dead-end pores:

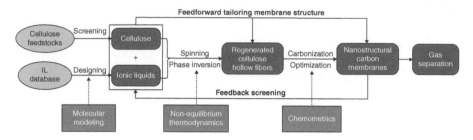

FIGURE 10.1
An advanced process for cellulose-based carbon membrane development.

A. Catalysis using Lewis acids (HCl) or ionic salts (NH_4Cl) can make carbon fibers mechanically stronger. For safety, the NH_4Cl will be put inside a quartz tube, and a sweep gas (helium) used to slowly release HCl when the carbonization temperature reaches 340 °C (i.e., the NH_4Cl decomposition temperature). Burst pressure testing and dynamic mechanical analysis will be conducted to document mechanical strength;

B. Bundling both sides of cellulose precursors to make straight carbon hollow fibers (both sides can slide when the precursor shrinks during the carbonization process);

C. Cooling the outside of the furnace so common epoxy glue can be used for sealing (<150 °C);

D. Controlling the micropore volume and pore size using an *in situ* gas (helium) permeation unit (tailoring pore size at the molecular level). The carbonization process can stop automatically when helium permeance reaches the setting target based on an online gas permeation measurement system. The carbonization system will be evacuated using a vacuum pump, and helium will permeate through the membrane and accumulate in a fixed volume at the permeate side during carbonization. Permeate pressure will increase over time;

E. Rotating the furnace during carbonization to make straight fibers and push chars/carbon ashes to the ends of the fibers by centrifugal force.

The proposed carbonization procedure may have a high risk related to the challenges of controlling gas permeation during carbonization (it wrongly indicates high permeate flow if some fibers are broken in step D) and modification of carbonization equipment (e.g., steps B and E). However, it has high gains in making mechanically strong, aging-free carbon membranes that are controlled at the molecular level (micropore volume and pore size tuned; beneficial for module making and real-life applications). This will provide a ground-breaking contribution to the preparation of frontier, high-performance carbon hollow fiber membranes for gas or liquid separations.

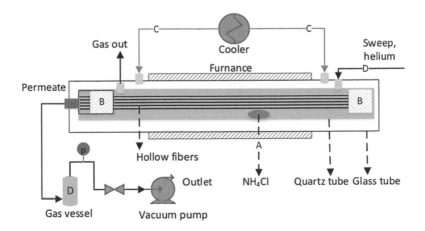

FIGURE 10.2
A novel carbonization process.

Moreover, carbon membrane module design and operation should be considered for upscaling. Lab-scale testing usually uses a shell-side feeding configuration, which may not be optimal when the system is operated at a high-pressure feed flow. In case of vacuum operation, feeding from the bottom may provide a better membrane performance due to gravity. The carbon membrane should be regenerated over time to recover performance because membranes deteriorate when exposed to water vapor or higher hydrocarbons, and proper pre-treatment units should be installed to protect carbon membranes and thus achieve a longer lifetime, which will reduce capital cost. Nevertheless, mechanically strong carbon membranes with high performance are promising for gas or liquid separation, especially for high-temperature and high-pressure separation processes. Hydrogen purification from syngas may become a major application for carbon membranes, but CO_2 removal from natural gas or biogas (CO_2/CH_4 separation) also has great potential. However, in order to bring carbon membrane technology to future commercialization, the above-mentioned challenges must be overcome.

Index

Page numbers in *italics* refer to figures and those in **bold** refer to tables.